NOBEL CO

MW01491103

# BANG:

## The Evolving Cosmos

**William A. Fowler**, Nobel Laureate, California Institute of Technology
**Timothy Ferris**, University of California, Berkeley
**Margaret Geller,** Harvard University
**Edward Harrison**, University of Massachusetts
**Ernan McMullin**, University of Notre Dame
**Philip Morrison**, Massachusetts Institute of Technology
**Richard Fuller, editor**, Gustavus Adolphus College

## Edited by
# Richard Fuller

Gustavus Adolphus College
Saint Peter, Minnesota 56082

Copyright © 1994 by
**Gustavus Adolphus College**

**University Press of America®, Inc.**
4720 Boston Way
Lanham, Maryland 20706

3 Henrietta Street
London WC2E 8LU England

**Library of Congress Cataloging-in-Publication Data**

Nobel Conference (27th : 1991 : Gustavus Adolphus College)
Bang : the evolving cosmos / Nobel Conference XXVII ; William
A. Fowler ... [et al.] ; Richard Fuller, editor.
p.     cm.
1. Cosmology—Congresses.   2. Olber's paradox—Congresses.
3. Cosmic bacground radiation—Congresses.   I. Fowler, William A.
II. Fuller, Richard M.   III. Title.
QB980.N63     1991     523.1—dc20     94–7368 CIP

ISBN 0–8191–9468–9 (cloth : alk. paper)
ISBN 0–8191–9469–7 (pbk. : alk. paper)

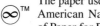 The paper used in this publication meets the minimum requirements of
American National Standard for Information Sciences—Permanence
of Paper for Printed Library Materials, ANSI Z39.48–1984.

# Contents

# Acknowledgments

The time was right, following Nobel Conference XXVI on Chaos: The New Science. It was most appropriate to have the topic of Nobel Conference XXVII be The Evolving Cosmos." In this conference we would explore the state of our knowledge of the oldest science. Cosmology. The conference was put together by Richard Elvee, the Director of the Nobel Conference series and the conference committee made up of Eric Eliason, Tom Huber, Michael Hvidsten, Colleen Jacks, Charles Niederriter, and Richard Fuller, Chair.

The success of this conference was due to contributions made by many other members of the Gustavus staff, faculty, and students. We are especially grateful to the work of the entire staff of the Public Affairs Office, the Office of Media Services, the Food Service, the Music Department and the faculty and student hosts for our speakers.

The posters and brochures for the conference were produced by Kelvin Miller of Primarius, Ltd. Most important for the preparation of this volume has been editing work of Elaine Brostrom and the secretarial services provided by Janine Genelin.

Finally, we acknowledge the funding that makes the annual Nobel Conferences possible. The major Nobel conference endowment gift by the late Russell and Rhoda Lund has been supplemented by grants from Pat Lund, the Cray Research Foundation and the Minnesota Humanities Commission, in cooperation with the National Endowment for the Humanities and the Minnesota State Legislature.

# INTRODUCTION

Richard Fuller

     The XXVII Nobel Conference at Gustavus Adolphus College was held on October 1 and 2, 1991. The subject of the conference was set forth in the invitation letter sent to our speakers as follows: One of the most useful and universal metaphors in all of science is that of evolution. It has become evident that the evolutionary model relates explanations for structures of the microworld of the quantum and elementary particles to the explanations for the content and structure of the universe itself. The main goal of this conference is to present the experimental data and theoretical framework that make the evolutionary model so compelling. We invite you to join us in the discussion and celebration of both the known and the mystery that make up our current comprehension of the evolution of the cosmos.

     The participants who accepted this invitation for the Nobel Conference were: Dr. Edward Harrison, Professor in the Five College Astronomy Department at the University of Massachusetts; Dr. Philip Morrison, Institute Professor Emeritus of Physics, Massachusetts Institute of Technology; Dr. Timothy Ferris, Professor of Journalism and Astronomy, University of California, Berkeley; Dr. William Fowler, Institute Professor Emeritus of Physics, California Institute of Technology; Dr. Margaret Geller, Professor of Astronomy and Astrophysics, Harvard University and Smithsonian Astrophysical Observatory; and Dr. Ernan McMullin, O'Hara Chair of Philosophy and Director of the program of History and Philosophy of Science at the University at Notre Dame.

     Dr. Harrison's opening lecture, "Have You Seen the Big Bang Lately," provides us with an interesting and provocative historical account of the cosmological implications of Olber's paradox—'Why is the sky dark at night?'

     Dr. Morrison followed with his lecture, "Newton and Anti-Newton; Enforced Simplicity, Inaccessible Origins." Here he describes the cosmic drama of three acts. We know the third act of our own time the best. This encompasses about twelve billion years. This act includes the birth and death of stars and the newly described structure in the universe that Dr. Geller reveals in her lecture. This is the act governed by the physics of Newton. There is an earlier act (second) of the inflationary universe that is ruled by Anti-Newton physics. Morrison points out that we are beginning to understand this act better as we expand our experimental data base. He tells us that at this time, the first act is inaccessible to us. Dr. Morrison leaves us with the question, are there other acts in this cosmic drama?

Dr. Ferris gave his lecture, "Evolution in Interstellar Communications Systems," as a challenge for us in coming to understand our place in the universe. He speculates about the search for extraterrestrial intelligence, SETI. Dr. Ferris points out what we know today and how we undertake this search. He describes for us the establishment of an interstellar communication network that he sees as a natural response to our privilege of existence.

Dr. Fowler's lecture, "Early Nucleosynthesis in an Inhomogeneous Universe," was an exposition of the beautiful theory which explains the nuclear synthesis that resulted in Dr. Fowler's Nobel Prize. Dr. Fowler raises the question about the "missing matter" that is necessary for the theory to hold. He is optimistic that we will find the answer to this very important cosmological question.

Dr. Geller's illustrated lecture, "Where the Galaxies Are," was an inspiring account of her research that has revealed an unexpected structure of the universe. Her maps of the galaxies provide us with an excellent example of the power of modern astronomy in providing us with basic knowledge about the evolution of the universe.

Dr. Eman McMullin's lecture, "Long Ago and Far Away: Cosmology as Extrapolation," gives us an elegant and lucid account of cosmological knowledge from its beginnings in Babylonia and Greece to the modern inflationary hypothesis. Professor McMullin provides us with an excellent framework to place this entire Nobel Conference in its proper historical context.

At the final Nobel Conference Dinner, our Nobel Conference Director Richard Elvee asked the panelists to respond to his questions "For what do you hope? What tools do you hope to get your hands on or see come into use? What theories disconfirmed?" We include in this volume the poignant responses from Professor Morrison and Professor McMullin. They challenge us as well as ennoble us for addressing the future of cosmology. A fitting ending for this volume.

Finally, we offer this volume to readers in the hospitable spirit which was so evident in the 27th Nobel Conference—welcome to the lectures, conversations, and insights concerning our evolving cosmos.

# Contributors

**William A. Fowler**
Institute professor emeritus of physics, California Institute of Technology; Legion d'Honneur (1989); first recipient of the William A. Fowler Award for Excellence and Distinguished Achievements in Physics, Ohio Section, American Physical Society (1986); Nobel Prize in Physics (1983); awarded National Medal of Science (1974); awarded Barnard Medal for Meritorious Service to Science (1965); member, honorary member or fellow of numerous learned societies, including the Royal Astronomical Society, Astronomical Society of the Pacific and the National Academy of Sciences.

**Timothy Ferris**
Professor of journalism and astronomy, University of California, Berkeley (1986- ); two-time recipient of the American Institute of Physics Prize; writer and narrator of **"The Creation of the Universe,"** an award-winning PBS production; author of five books, including **Coming of Age in the Milky Way** (nominated for a 1989 Pulitzer prize) and **Galaxies** (nominated for a National Book Award in 1981); producer of the phonograph record carried aboard the Voyager interstellar spacecraft.

**Margaret Geller**
Professor of astronomy, Harvard University (1988-); astrophysicist, Smithsonian Astrophysical Observatory (1983- ); senior visiting fellow, Institute of Astronomy, Cambridge University (1978-1980); awarded AAAS-Newcomb Cleveland Prize (1991); elected to American Academy of Arts and Sciences (1990); author of numerous professional and popular articles, including **Research Frontiers in Astronomy** (1984), **The Universe Nearby** (1988), **Mapping the Universe: Slices and Bubbles** (1988), and **Surveying the Universe** (in preparation).

**Edward Harrison**
Distinguished university professor of physics, University of Massachusetts (1987- ); Melville S. Green Lecturer, Temple University (1988); recipient of Melcher Award (1986); fellow or member of several learned societies, including the American Physical Society, American Association for the Advancement of Science and the International Astronomical Union; author of many articles and papers published in professional journals; author of four books, including **Darkness at Night: The History of a Cosmological Riddle** (1986); and **Masks of the Universe** (1985).

## Ernan McMullin

O'Hara chair of philosophy and director, program of history and philosophy in science, University of Notre Dame (1984- ); fellow, International Academy of the History of Science (1988); recipient of Aquinas Medal, American Catholic Philosophical Association (1981); chairman, U.S. National Committee of the International Union of History and Philosophy of Science (1982-1984, 1986-1988); editorial board member or consultant to several professional journals, including the **British Journal for the Philosophy of Science** and **Astronomy Quarterly**; author of 10 books, including **Evolution and Creation** (1985) and **Rationality, Realism and the Growth of Knowledge** (in preparation).

## Philip Morrison

Institute professor emeritus of physics, Massachusetts Institute of Technology; AAAS-Westinghouse Public Understanding of Science Award (1988); Andrew Gemant Award, American Institute of Physics (1987); Public Education Medal, Science Museum of Minnesota (1982); chairman, Federation of American Scientists (1973-1976); book reviews editor, **Scientific American** (1965- ); member of Manhattan Project, University of Chicago (1942-1944) and Los Alamos Laboratory (1944-46), participating in first desert test of atomic bomb; author of several books and learned essays, including **Winding Down: The Price of Defense** (1979) and **Search for the Universal Ancestors** (1986).

# OUR EVOLVING VIEW
# OF THE UNIVERSE

## EDWARD HARRISON

The puzzle of cosmic darkness, nowadays known as Olbers' paradox, is the old riddle that asks, "Why is the sky dark at night?" Here I discuss how the riddle has changed over the centuries with our evolving view of the universe. The riddle began in Elizabethan England soon after the Copernican revolution and has since been influential in shaping the history of astronomy and cosmology. In the twentieth century we have discovered that the universe is expanding, is of finite age, and in the beginning was extremely dense and hot. The riddle climaxes with the realization that the entire sky is covered not by distant unseen stars but by the big bang redshifted into an infrared gloom.

THE EMPYREAN

Figure 1: The "Empyrean" by Gustav Dore, showing Dante and Beatrice gazing on the blazing light of the angelic spheres.

Figure 2: In an unbounded universe, endlessly populated with stars, every line of sight should ultimately intercept the surface of a star. Why then is the whole sky not covered with stars? Why is the sky dark at night? This is known as Olbers' paradox. Wilhelm Olbers in 1823 was the first to use the line-of-sight argument. (From *Darkness at Night* by the author.)

Throughout the Judeo-Christian-Moslem world until Elizabethan times the idea prevailed that the twinkling stars all existed at the same distance from the Earth. They formed an enclosing "sphere of stars." Beyond the sphere of stars in the medieval universe stretched an extramundane and mysterious realm—the empyrean or heaven—a place where God dwelt. To a person ascending the celestial spheres (as narrated in *The Divine Comedy* by Dante), the blue light of the spheres grew brighter, and the empyrean, blazing with light, was the realm of purest fire (Figure 1). Quite naturally, in the sublunar sphere, the sky at night was dark because the heavenly light could not be seen by mortal eyes. The riddle of cosmic darkness had yet to be born. In this essay, I draw on the history of the dark night-sky riddle to illustrate how our understanding of the universe has changed.

\*   \*   \*

The riddle of why the sky at night is dark emerged in its simplest form in 1576 when Thomas Digges of London, an astronomer and a mathematician, proposed in *The Perfect Description of the Celestial Orbs* that stars are "fixed infinitely up," as in Figure 2. He dismantled the sphere of stars and dispersed all the stars throughout the endless space of heaven. Why did the now innumerable scattered stars not make the night sky bright? Because, said Digges, most stars are too far away to be seen. Digges originated the riddle of cosmic darkness in a rudimentary form, and he solved the riddle in a rudimentary and commonsense way: although space contains innumerable stars stretching away without limit, the sky is not ablaze with starlight because most stars are too far away to be seen individually.

Other astronomers in the sixteenth and seventeenth centuries came to much the same conclusion. But this apparently sensible solution of the riddle fails on closer examination. The excited atoms in a candle flame cannot be seen individually, but their light emitted collectively accounts for the brightness of the flame that we see. A similar argument applies to a multitude of luminous stars; although we cannot see them individually, their collective or fused light should create a sky that shines bright at every point of the sky. Digges, of course, was not aware of excited atoms, but other and simpler analogies exist. The Roman poet Lucretius in his epic poem *De Rerum Natura* (about 55 B.C.) wrote that the flashes of bronze of maneuvering legions, seen from a distance, fuse into "a blaze of light stationary upon the plain."

Astronomers have stated the riddle in many different ways. The most lucid and compelling statement was made in 1823 by the German astronomer Wilhelm Olbers:

Space has no edge and is therefore unbounded. Suppose that unbounded space is populated endlessly with stars. A line of sight, as in Figure 3, in any direction from the eye, extending away from the Earth into the depths of space, eventually must intercept the surface of a star. Hence, in all directions and at all points the entire sky should be covered in stars with no separating dark gaps. If we suppose that most stars are bright like the Sun, then every part of the sky must shine as bright as the Sun's disk. But fortunately for astronomy, said Olbers, the sky is quite dark, particularly on moonless nights. This head-on collision between compelling theory and indisputable observation is the essence of the riddle.

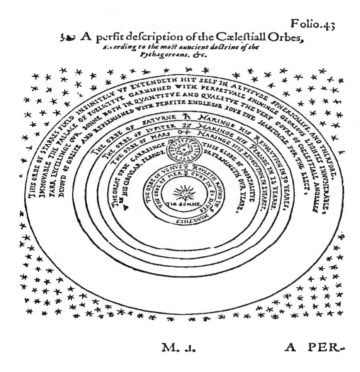

Figure 3: Thomas Digges' cosmographical diagram in "The Perfit Description of the Caelistiall Orbes" (1576). The sphere of stars has been dissolved and the individual stars have been dispersed throughout infinite space.

For two hundred years after *The Perfect Description*, records show that almost every astronomer and every scientist interested in cosmology made some contribution to the riddle of cosmic darkness. Edmund Halley of comet fame, a man of great originality, in 1721 discussed the infinity of the universe and the possible consequences of a universe populated endlessly with bright stars. He introduced the idea of constructing large imaginary concentric shells of constant thickness (see Figure 4), with the observer located at the center. Halley assumed that stars are everywhere similar to the Sun and distributed uniformly in space. The volume of any shell, he argued, is proportional to the square of its radius; the number of stars in the shell is proportional to its volume, and hence to the square of the radius. In this way, Halley started on the right track, but he then made a false turn in his argument and came to the same conclusion as Digges: the sky is dark because the beams from individual stars "are not sufficient to move our sense."

Exactly what Halley did is not clear. He should have argued that the light received from a star in any shell is proportional to the inverse square of its distance, and therefore each shell contributes an equal amount of light. For example, if we consider two shells, one double the radius of the other, the larger shell contains four times as many stars, but each star in this larger shell gives a quarter of the light given by a star in the smaller shell.

Halley also said the same argument applies to the disks of stars, and (after correcting his argument) we find that shells contribute equally to covering the sky with stars. If we again consider two shells, one double the radius and containing four times as many stars as the other, each star in this larger shell has an apparent geometric size a quarter of the size of a star in the smaller shell.

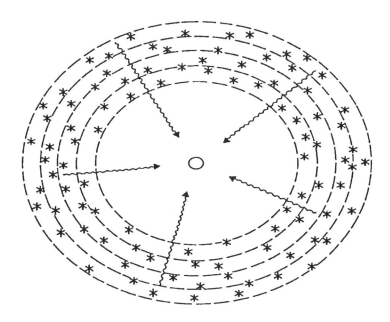

Figure 4: Edmund Halley's shells. Large imaginary concentric spheres are constructed with the observer at the center. The spheres form shells of constant thickness, as shown. The volume of each shell is proportional to the radius squared, and the number of stars in each shell is therefore also proportional to the radius squared. But the light received at the center from each star is inversely proportional to the radius squared. Hence, all shells contribute equal amounts of light.

A few years later in 1744, the young Swiss mathematician and astronomer Jean Phillipe Loys de Cheseaux did not make the same mistake. He was the first to perform the correct calculations. He adopted Halley's method of constructing imaginary concentric shells of stars, and showed that each shell contributes equally to the amount of the sky covered by the disks of stars. He added shells out to a distance of 3 thousand trillion ($3 \times 10^{15}$) light years (in modern units of distance), and at this distance the sky became fully covered by star disks. Stars farther away than this distance could not be seen because they were occulted by nearer stars. Cheseaux's calculations showed that the total number of stars covering the sky was roughly $1 \times 10^{46}$. The whole sky is 180,000 times larger than the Sun's disk, said Cheseaux, and if every point of the sky shines as bright as any point on the Sun's disk, the total starlight falling on Earth should be 180,000 times more intense than sunlight. "The enormous difference between this conclusion and experience," said Cheseaux, led him to suggest that an interstellar absorbing medium attenuates starlight as it travels from the multitudes of distant stars.

Most modern accounts of Olbers paradox, popular and scholarly, discuss the amount of radiation received from Halley's shells, but not the amount of the sky covered by stars. Often these modern accounts wrongly conclude that because each shell contributes an equal quantity of starlight, and because an infinity of shells exist in an infinite universe, the starlight must be infinitely intense. This conclusion applies at every point in space, and starlight must therefore be infinitely intense everywhere. To my knowledge, no astronomer before the twentieth century ever came to such a fallacious conclusion. Most astronomers looked at the riddle in terms of the sky-cover argument, and obviously one cannot add an infinite number of shells when considering how many shells are required to cover the sky with stars. Remarkably, the sky-cover argument has been largely forgotten in the twentieth century, despite the popularity of Olbers' paradox. Stanley Jaki, a historian of science, in his, book *The Paradox of Olbers' Paradox*, states forcefully that a paradoxical characteristic of Olbers' paradox is the failure by scientists to consult the historical records. One of the paradoxical features of this book is the author's failure to realize that pre-twentieth century astronomers treated the riddle as a simple sky-cover problem, and not as a complex problem in radiation theory. Had Jaki as a historian noticed the vast difference between past and present arguments, he might have modified his strong belief that the radiation intensity is infinitely great in an infinite universe.

*   *   *

Olbers in 1823, in an article "On the transparency of space," discussed the riddle of cosmic darkness in some detail. His treatment followed in some ways that previously given by Cheseaux and his solution was also the absorption of starlight in interstellar space. But he did not calculate the distance to the background of stars, or the number of stars needed to cover the sky, or the intensity of starlight from a star-covered sky. He used Halley's shells and also introduced the line-sight-argument to show that even in a universe that is not uniformly populated with stars, the sky would still be covered with stars.

The forest analogy (see Figure 5), not used by Olbers, makes clear his line-of-sight argument. In a large forest of tall and well-spaced trees, every horizontal line of sight from the eye terminates at a tree trunk. The visible trees overlap one another and fuse into a distant background that surrounds a person like a circular wall. Nearer than the background distance stand visible or partly visible trees, and beyond the background distance stand invisible trees. The distance to the background is simply the square of the average separating distance between trees divided by the diameter of a typical tree. Similarly, the background distance of the stars is the cube of the average separating distance between stars divided by the cross-section of a typical star such as the Sun. A misty forest, in which only the foreground trees are visible through the mist, illustrates the starlight absorption solution given by Cheseaux and Olbers.

Figure 5: How far can we see in a forest? Or how far can an arrow travel? The background distance--the average distance before intercepting a tree trunk is the area occupied by a single tree divided by the width of a tree trunk. If trees have an average separating distance of 10 meters, and the width of a trunk is half a meter, the background distance is 200 meters.

Although the absorption solution seems plausible, only five years after Olbers published his solution, the famed astronomer John Herschel proved that absorption cannot be the answer. To understand Herschel's argument, we must try and imagine a universe full of starlight 10,000 billion times brighter than present starlight. In this intense inferno of radiation, much like falling into the Sun, the whole Earth would turn to vapor in a few hours. Clearly, any absorbing medium in interstellar space, such as gas and dust, would heat up rapidly and cease to absorb, and would then merely diffuse the radiation. Since Herschel's devastating argument, no person except Edward Fournier d'Albe has suggested that absorption solves the riddle of cosmic darkness. Fournier d'Albe, a British engineer who struggled to heal Anglo-Irish relations, suggested in the early years of this century that stars cover the sky but most of them are nonluminous. These nonluminous stars intercept and absorb the light of luminous stars.

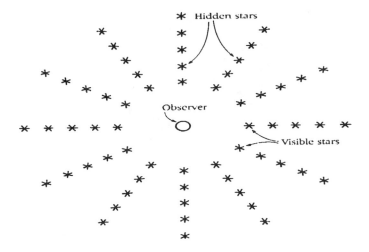

Figure 6: An improbable solution of why the sky is dark at night, proposed in jest by Edward Fournier d'Albe. The stars hide behind one another in rows, he said, thus making the sky dark at night. Notice that Fournier d'Albe has cheated; each shell of stars contains the same number of stars and therefore the light from each shell decreases as the inverse square of distance. With such a distribution, the sky remains dark even if the stars did not hide behind one another. (From *Darkness at Night* by the author.)

His absorption theory could work, but is totally inconsistent with modern observations. Fournier d'Albe proposed several solutions of the riddle, and jokingly he suggested that the sky is perhaps dark because stars hide behind one another, as illustrated in Figure 6.

*    *    *

Over the centuries numerous solutions of the riddle have been proposed. Broadly speaking, they divide into two groups, and each group is based on a different interpretation of the state of darkness of the night sky. The first group accepts interpretation A (shown in Figure 7) that visible and invisible stars fully cover the sky. The stars are all there, out to the background, but the starlight from the very distant stars is missing and we see only nearer stars. Most discussions of the riddle in recent decades have adopted this interpretation. The question, "why is the sky dark at night?" becomes "where has all the starlight gone?"

INTERPRETATION A

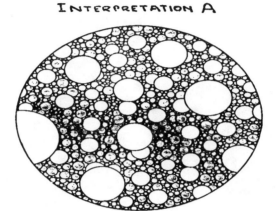

Figure 7: Interpretation A: The sky at every point is covered with stars, most of which cannot be seen. Where has the missing starlight gone? (From *Darkness at Night* by the author.)

A second group of solutions, less recognized in recent times, accepts interpretation B (shown in Figure 8) that all the stars of the universe fail to cover the whole sky. The question, "why is the sky dark at night?" becomes "where have all the stars gone?"

INTERPRETATION B

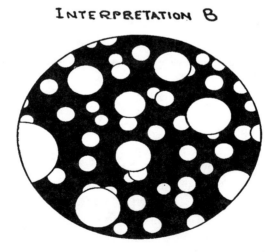

Figure 8: Interpretation B: The sky is not covered with stars. What has happened to the missing stars? (From *Darkness at Night* by the author.)

Digges, Halley, Cheseaux, Olbers and most astronomers before the second half of the nineteenth century accepted interpretation A. Interpretation B, however, became increasingly popular in the nineteenth century and was indeed the interpretation adopted in the Victorian universe as enunciated by Agnes Clerke in 1890:

> But the probability amounts almost to certainty that star-strewn space is of measurable dimensions. For from innumerable stars a limitless sum-total of radiations should be derived, by which darkness would be banished form our skies; and the "intense inane," glowing with the mingled beams of suns individually indistinguishable, would bewilder our feeble senses with its monotonous splendor.

The standard model of the universe at the end of the nineteenth century consisted of an immense collection of stars, the Milky Way or Galaxy, with the Sun near the center, and beyond the Galaxy stretched an endless dark void. The stars in the Galaxy, though numerous, were insufficient to cover the whole sky, as in interpretation B. The dark night sky gave immediate proof of the endless starless void beyond the Galaxy. "With the infinite possibilities beyond," said Clerke, "science has no concern." Serious scientists at that time avoided discussing the distant universe; this was a very sensitive subject that had theological overtones inherited from the medieval universe.

The history of the riddle contains numerous discussions and proposed solutions but few calculations. Mathematical analyses, when they occur, come like a refreshing wind sweeping away the fog of unfounded opinion. Lord Kelvin in a paper entitled "On ether and gravitational matter through infinite space," published in the *Philosophical Magazine* in 1901, at last performed the definitive calculations. He showed, in the context of the standard model of his day, that interpretation B was correct and the Galaxy contained insufficient stars to cover the sky. But, Kelvin said, suppose that stars stretched away to much greater distances, perhaps even filling an endless universe; the visible stars would still fail to cover the whole sky. Kelvin's explanation is as follows.

Light travels at finite speed, and when we look out in space we look back in time. We see the Sun as it was 500 seconds ago, and the nearest stars as they were roughly five years ago. Kelvin had shown in previous work that stars cannot shine endlessly, pouring out energy in the form of starlight, but must have only a finite lifetime determined by their energy resources. We cannot look out in space and see stars at infinite distance because that means looking back across an infinite span of time to when luminous stars did not exist. When we look out in space, and look back in time a period equal to and greater than the age of luminous stars, we see the darkness that existed before the birth of luminous stars. Kelvin showed that even at this great distance, there are still insufficient stars to cover the entire sky, and the darkness we see between the distant stars at night is actually the darkness that existed before the birth of stars. Thus, interpretation B applies even in an infinite universe. To the question, Where is the missing starlight? The answer is it has not yet reached us.

Edgar Allan Poe in 1840 suggested a similar answer. In his cosmological essay *Eureka* he wrote,

> Were the succession of stars endless, then the background of the sky would present us an uniform luminosity, like that displayed by the Galaxy—*since there could be absolutely no point in all that background at which would not exist a star.* The only mode, therefore, in which, under such a state of affairs, we could comprehend the *voids* which our telescopes find in innumerable directions, would be by supposing that the distance of the invisible background so immense that no ray from it has yet been able to reach us at all.

Poe, an essayist and poet, lacked the mathematical skill needed to show that his ideas was feasible. Ole Roemer in 1676 discovered the finite speed of light, and many astronomers since Roemer, including Cheseaux, had the skill but not the idea. The habit of thinking of astronomical distances in terms of light-travel time was slow to develop. Also, the realization that when we look out in space we also look far back in time to the creation of the universe trespassed on theological territory and for this reason the idea was probably not discussed openly.

$$* \quad * \quad *$$

Modern estimates show that the background distances of the stars is $10^{23}$ light years; stars must stretch away to this enormous distance if they are to cover all the sky. This means, to see a fused background of stars at this immense distance, the stars must have been shining for at least $10^{23}$ years.

But a star like the Sun has a luminous lifetime of $10^{10}$ years, which we can take as a sort of average lifetime of stars in general, and therefore, if all stars commence shining at about the same time, we shall see them out to a distance of only $10^{10}$ light years. Kelvin estimated that starlight intensity is about one ten-trillionth ($10^{-13}$) of the intensity anticipated by the riddle, and coincidentally his result agrees with modern estimates. Kelvin showed also that only one ten-trillionth of the sky is covered geometrically by the disks of stars.

Lord Kelvin was the first and only person to relate mathematically the intensity of starlight and the sky-cover fraction, thus demonstrating the aptness of Olbers' geometric line-of-sight argument. In view of the great interest shown in Olbers' paradox in recent decades, it is surprising that Kelvin's work was undiscussed until 1986. Also, writers who have consulted Olbers' work on the intensity of starlight and have fostered interest in the riddle of cosmic darkness have unaccountably failed to note his line-of-sight argument. This is unfortunate because this elegant demonstration of the riddle is much easier for most people to understand than Halley's shell argument, and it has the advantage that we avoid the temptation of concluding that the intensity of starlight must be infinite in a universe of infinite extent.

\*   \*   \*

Our understanding of the universe and the meaning of the word universe has changed considerably since the time of Lord Kelvin. Giant telescopes of great precision have surveyed the skies and probed deep into space. In the early years of the twentieth century the "great debate" flared up among astronomers: did the universe contain just one Milky Way (or Galaxy) surrounded by an infinite void, or did the universe consist of multitudes of milky ways (or galaxies) scattered in endless space? The debate ended in the 1920s and the many-galaxy universe prevailed.

Albert Einstein's theory of general relativity reached its final form in 1916. In the following years telescopic observations showed that the galaxies were moving away, indicating that the astronomical universe was expanding. Various versions of an expanding universe emerged, and by the middle of the century the two main contenders were the big bang universe of finite age first proposed by Alexander Friedmann of Russia and Georges Lemaitre of Belgium, and the steady state universe of infinite age later proposed by Hermann Bondi, Thomas Gold, and Fred Hoyle in 1948.

Cosmology at that time offered a choice between the instant creation of a big bang and the little-by-little creation of a steady state. To many scientists the steady state version seemed more respectable because it avoided entanglement with the creation scriptures. Not surprisingly, Friedmann dabbled in big-bang cosmology shortly after the violent and explosive Bolshevik revolution. Lemaitre spent a lot of time thinking about the creation and early stages of the universe, but that was not surprising, he was a priest. George Gamow led an investigation into the nature of the early universe, but many regarded him as a slightly crazy Russian, and largely ignored this important aspect of his work.

Hermann Bondi revived the riddle of cosmic darkness in the context of the steady state universe and attributed the darkness of the night sky to the expansion of the universe. The wavelengths of light traveling in space are continuously stretched as the universe expands, as shown in Figure 9.

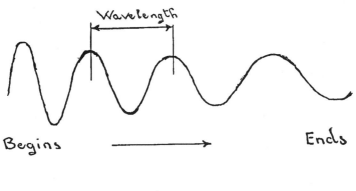

Figure 9: Wavelength stretching. A wave of radiation is progressively stretched while traveling in an expanding universe.

This stretching moves the emitted light toward the red end of the spectrum and accounts for what is referred to as *redshift*. Bondi argued that the redshift effect is sufficient to move the light of very distant stars out of the visible region of the spectrum into the invisible infrared region. This explanation assumed that the entire sky is covered with stars, as in interpretation A, and the missing starlight actually exists but has become invisible to the human eye.

Bondi's expansion redshift explanation had wide appeal. It had also a sensational element, for the darkness of the night sky gave us all direct proof that the universe is expanding. No cosmological observation could be simpler and no conclusion more dramatic. The redshift explanation was applied to all expanding universes. Darkness proved expansion. Kelvin's argument, showing that the sky is not covered in stars in a universe of finite age, was forgotten. Subsequent calculations showed that Bondi's redshift explanation was correct only in the steady state universe and not in the big bang universes of finite age. In retrospect we realize that if the night sky is dark in a static universe of finite age, as shown by Kelvin, the night sky is even darker in an expanding universe because of the redshift effect.

* * *

The discovery by Arno Penzias and Robert Wilson in 1965 of the cosmic background radiation ended the controversy and made clear that we live in a big bang universe that originated in a very dense and hot state. The observed cosmic background radiation is nothing less than the afterglow of the big bang. Because of this awesome discovery, the early universe became scientifically respectable and a fit subject for serious study.

Something was seriously wrong with Olbers' paradox. If we imagine that all matter in the universe is annihilated and the released energy $E=mc^2$ converted into thermal radiation, the night sky would still be dark. Our universe at present does not contain enough energy to create the bright sky visualized in Olbers' paradox. Calculations showed that the intensity of starlight in a big bang universe, such as our own, must be very low, not because of the redshift effect, but because of the limited age of the universe. Stars have been luminous for only 10 billion years, and not enough starlight has been emitted to make the night sky bright. In other words, the sky is not covered by stellar disks, and interpretation B is correct. Kelvin's argument is sufficient, for if the sky is dark in a static universe, it is also dark in an expanding universe of similar age and constitution.

* * *

A line of sight from Earth stretches out in space and back in time. The line-of-sight cannot extend to infinite distance in a universe of finite age; instead, it reaches out and back to the beginning of the universe, as in Figure 10.

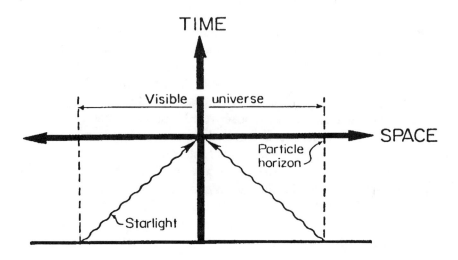

Figure 10: This diagram shows that as we look out in space we also look back in time. Rays of light travel to us from all parts of the sky, and from as far back as the beginning of the universe. What happened in the beginning covers the whole sky.

In the nonexpanding universes of the past, from Newton to Kelvin, the beginning was entirely a theological subject, wisely avoided by creditable members of society. Most astronomers avoided drawing attention to the fact that when we gaze at distant stars we look back millions of years in time. The immensity of the visible heavens and the finite speed of light contradicted scriptural testimony, and only radicals, such as Mark Twain in his posthumously published *Letters From Earth*, dared to point out that if the heavens were created a few thousand years ago, we should see only a comparatively small number of stars within a distance of a few thousand light years. According to interpretation B, we look out between the stars back to the creation of the universe.

\*   \*   \*

In the 20th century we have become acutely aware that when we look out in space we also look back in time. In the farthest depths of space we see the universe as it was long ago before the birth of stars and galaxies. In the expanding universe we look back to the big bang (Figure 11). We cannot see the beginning, the creation, for that lies veiled by the incandescent light of the early universe. The light that was once intensely bright has been dimmed by the expansion of the universe and survives as the cosmic background radiation.

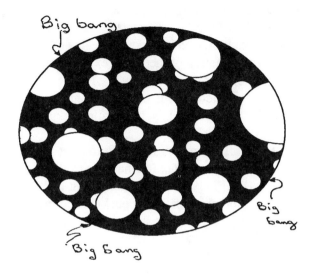

Figure 11: The big bang, looming in the dark gaps between the stars and galaxies, covers the whole sky. The big band was once intensely bright, but is now redshifted into the infrared by the expansion of the universe and is invisible to the eye. (From <u>Darkness at Night</u> by the author.)

The four hundred year-old riddle of cosmic darkness climaxes in this century with the startling realization that the night sky is covered not by multitudes of unseen stars, but by the big bang of the early universe. Every line of sight, if not intercepted, terminates in the big bang. The intensely bright light of the big bang, which covers the entire sky, has mercifully been redshifted into an invisible infrared gloom.

We have apparently arrived at a fitting conclusion to a riddle that has teased astronomers and cosmologists for centuries. But wait! History shows that the riddle constantly changes with our evolving view of the universe. I am quite sure that we have not seen the end of Olbers' paradox.

References may be found in the author's book *Darkness at Night: A Riddle of the Universe* (Harvard University Press, Cambridge, Massachusetts, 1987).

# Questions and Comments

**Ferris:** I'd like to ask Professor Harrison if he would care to say a few words about the results of the COBE satellite and its observations of the cosmic background radiation which marked, it seems to me, such a remarkable chapter in the confirmation of theory by experiment.

**Harrison:** What do you want to know?

**Ferris:** Given the title of your talk, I thought you might want to comment about the results obtained by the cosmic background satellite and its relevance to the Big Bang Theory.

**Harrison:** Well, this thermal radiation in the last year or so has been discovered to be remarkably isotopic. It's the same in all directions to within about 1 point in 100,000. This, after subtracting out the fact that the solar system is in motion in the universe. This background radiation gives us a preferred space that's homogeneous and isotropic in which once again absolute motion has been restored if we subtract out our motion of about 600 kilometers a second from it. That is a puzzle. It agrees with the fact that one might expect the early universe to be isotropic but it is in conflict to some degree that optically the universe that is emergent, is highly structured. Margaret Geller knows more about that than I do. Structure is evident in the universe on the scale of galaxies, clusters of galaxies, superclusters, great voids. How is it possible that an expanding universe that forms this structure can emerge out of a dense radiation field and leave that radiation field without almost any irregularity at all? You'd expect some irregularity in the cosmic radiation that mimics or relates to the fact that optically the universe is highly structured.

**Morrison:** I have a question of more historical nature which I was fascinated by the ascription of an early recognition of the time that buries the infinite star sphere if it exists by Mark Twain. And I seem to recall also that Edgar Allen Poe had an early insight about this. Do you believe that that is the case?

**Harrison:** Mark Twain in his posthumously published letters from earth mentioned that on the day of creation there would be no stars. You'd have to wait a few years for the first star to emerge and then more and more stars would cover the sky. His relatives and trustees repressed this heretical work that didn't fit in with the sedate society of Connecticut. This really does indicate how cosmology is influenced very strongly by fashions of thought. Edgar Allen Poe was another radical not accepted by the establishment and he also in his essay, Eureka, a year before he died, pointed out that the darkness of the night sky might be because the universe is not old enough for the light from distant stars to reach us. But you see, the history of this puzzle is full of these surmises and it's rare that you find a proper calculation. I tried to keep it to those people that did those calculations, there are lots of opinions, but very few calculations. Cheseaux and Lord Calvin calculations are really the two champions in this subject.

**Fuller:** We have a question from the audience. From the hypothesis discussed, do you feel that the universe is headed for the entropy death?

**Harrison:** I wish the speaker would tell me what he or she means by entropy, then I would know on what level to answer this question. But, in the last century it was feared that universe would eventually suffer a heat death. Later that was then expressed in terms of an entropy death. But the more fearsome, entropy is by and large conserved and all the entropy is in the microwave radiation. Stars pouring their hearts out in starlight will never automatch the entropy of the universe. It's all stored in the microwave background radiation. No, what is more fearsome nowadays is not the heat death, it's the energy death of the universe. Because Freeman Dyson has looked ahead a long way and when all the stars have died and when all forms of accessible energy have disappeared, how is life going to survive in the universe that expands forever getting darker and darker? And Freeman Dyson's answer to this is that, okay, let's slow up life. Living a hundred years in the universe, ten billion years old, is no different from living a million years at 10,000 times slower in a universe that's—oh I can't do the calculation—that is 10,000 billion years old or something. And you can go into long periods of hibernation so you husband or preserve these diminishing energy resources. I think all this is rather missing the point. Human beings have risen to the level of intelligence in the last million years. Can we imagine what if human beings survive, what level of intelligence they will achieve in another million years? And a million years is nothing on the time scale of the universe. What level intelligence will retain in a billion years time or in a hundred billion years time? I personally think that intelligent life will have taken over the control of the universe. It would not be trying to leak out slender reserves of energy. Once you harness the expansion of the universe, the ultimate form of unlimited energy, then there's no need to worry about an energy death. Fate of intelligent life in the universe has surely to take over control of the universe.

**Fowler:** Back on the question of the fact that the cosmic background radiation is so uniform across the sky, the observations say it's something like at most one part in ten to the fifth. In your point of view, Ed, what is the ultimate limit in your way of thinking? Is it one part in ten to the sixth? You see those of us who believe in the inhomogeneous universe are in real trouble because we can only get to ten to the minus four. But in your way of looking at things, how low can you go?

**Harrison:** Well, it's almost close to that limit within a factor of 2 or 3 and there is, even then that's slimming down the theoretical options. It's a matter of anxiety among theoreticians.

**Fowler:** So if the observers go any further, even in your point of view, you'll be in trouble?

**Harrison:** Yes, and there rests the reason for thinking the standard model of 100 years time might be different.

**Morrison:** I'm a little skeptical about this. Of course we're not in possession of all the theoretical structure to fill in what we don't know between the time of the recombination and the cooling of the microwave; the recombination of the atoms that are giving off the microwave into neutral transparent material and the time we can see that Professor Geller and the optical people see so well with all this rich structure of groups and voids, there's a lot to be filled in. We have strong reason to believe; not definite, but strong reason, there's a lot of matter we don't even know about, which matter can change, which can be of several types. I would say that I would not give up on this general picture, and I'll say that for almost any degree of isotopic. The more isotopic it is, the better I like it because what I say is new processes will be needed to make the galaxies and their distribution, but they're indicated strongly by the fact, the probability, that we don't see all the matter. We don't even know what kind of matter it is. How are we going to calculate what kind of galaxies are made out of stuff that we don't know about at all? Can't even give a name for sure, which can even change while it's going on? So I think that it's very premature to take the worry but it does show one very strong thing which I'm going to emphasize, that there was a uniform time and now there is an unstable gravitational coagulation time and both of those things are to be expected from what we know. And that's very reasonable. Now, that's it all filled out, that we have the last story, no, I don't know. But I strongly suspect that 100 years from now we're debating this kind of thing and not the general picture of the isotropy and the coagulation.

**Geller:** I'd like to just make a comment that I agree with Philip Morrison's view that it's not the global Big Bang picture that we have trouble with. It's making the structure, making galaxies, making out how the furniture of the universe essentially originates. The problem that we have is that we have a picture of the universe when it was about 100,000 years old. That's what this microwave radiation shows us. It's a picture. It's the photons from this epoch when the radiation was last scattered and it tells us that the early universe was very smooth and as I'll show you tomorrow, we have a picture of what the universe is like today, 10 or 20 billion years later. So, it's like going into a movie theatre at the beginning of the movie, then you fall asleep and you suddenly wake up at the end and you wonder what in the world happened in between and there's an awful lot that can go on in between and we don't really know all the things that might have happened in between. Fortunately, one of the things that's happening now is that large telescopes are being built on the ground and as Professor Harrison said, one of the wonderful things about the universe is that it's a sort of time machine. When you look out into space, you look back in time. So, if we can't figure out by pure thought how the objects of the universe got to be here, we'll observe it one day.

**Fuller:** We have a couple of questions and I wondered how long it would take before the big black holes got into the picture but we have two questions involving black holes. After the Big Bang, when and how did black holes develop? Anybody want to address that one?

**Geller:** Where are they?

**Morrison:** I'll believe in a black hole when I see one!

**Fuller:** One of the comments was, couldn't the darkness of the farthest stars be caused by black holes sucking the light out of the universe?

**Harrison:** The same person who made the jest about stars lining up one behind the other was Edward Fournier d'Albe. He was an engineer who broadcast television pictures from London in 1920 and he devoted his life to trying to heal Anglo-Irish relations and he proposed many solutions of this riddle in a wonderful book, small book, published in 1907. And he said one possibility is that perhaps only one star in 10 billion is luminous. All the rest are dark and so when we look out at nighttime at the sky, it is indeed covered with stars, but most of them are not luminous.

Now, you see, in a sense there is the nonluminous stars blocking the radiation and we can't see more luminous stars. It's a background of nonluminous stars, that is sucking radiation if you like, always blocking, absorbing the radiation from other stars. But, of course, even in his day he knew that was an unlikely explanation.

**Fuller:** We have another question. During the inflationary period, did the universe expand at a rate that exceeded the speed of light?

**Harrison:** Oh certainly, certainly. If I'm with you, I don't want to get launched on this subject because you see the expansion of the universe increases linearly with distance and so if the universe is infinite in extent, then bodies at infinite distance are receding from us at infinite speed. What one has to realize that we're now in the framework of general relativity and space has become not only curved but it's also become dynamic. It's become part of the physical universe. If you're going to create the universe, you've got to create space with it. It's part of the physical scheme of things and when the universe expands, it's not bodies moving through space. In that case they will be limited by special relativity. The motion light speed will be the limit, but it's the expansion of space and bodies are stationary in space. And it's the expansion of space that whops everything apart. The laws of expanding space are governed by general relativity and not by special relativity. So, 10 billion light years out from here space is receding at the speed of light and objects beyond there are receding faster and objects nearer are receding lessly. And a ray of light traveling toward us at the distance of 10 billion light years is actually stationary. It hurries toward us through space but space is itself expanding and carrying it away. And so beyond 10 billion years, light traveling toward us, in fact, is actually receding.

**Fuller:** We have one more question from the audience. How do you account for the discontinuous secretion of matter from a smooth background?

**Geller:** I have a note here which says: The movie wasn't so hot. It didn't have much of a plot. We fell asleep; our goose is cooked. Our reputation is shot. Wake up little Susie. From Don and Phil Everly. I guess this came from Tim. See, I'm naive. I wasn't educated about this; but it came from Tim Ferris.

**Ferris:** These are the words from an Everly Brothers song. The question was, how do you make galaxies out of a homogeneous smooth isotopic energy?

**Geller:** Well, if I knew the answer to that, I wouldn't be sitting here. I'd be sitting home writing it up. I think that's one of the fundamental questions. I mean, that's what I was saying before, that what we really don't understand is how the objects in the universe originate and it's a very difficult problem. The real conflict that we have is that we observe this very smooth early universe and now we observe a universe which is highly structured. The current models with what we think are reasonable assumptions, don't yield a match at the beginning and end of the movie. And, so something is missing from our understanding of how structure is made. But there's lots of room, as Phil said, for lots of physics to go on and we're missing a piece. I mean, just because we haven't thought of it doesn't mean that nature didn't.

# NEWTON AND ANTI-NEWTON: ENFORCED SIMPLICITY, INACCESSIBLE ORIGINS

## PHILIP MORRISON

It is plain that we human beings, reasoning, curious, believing creatures that we are, play some kind of a walk-on part in the great drama of the cosmos. Every culture of which we have some record has tried to grasp the problems of where we came from, where physically the entire world came from, how it was built—how its pillars were founded—and what would become of it. Of course we are no exception.

Professor Harrison's review of the history of cosmological ideas is somewhat daunting to a cosmologist of today, who hopes he is a little closer to things than some of our able predecessors were. There's no guarantee; I would accept that proposition. But there is hope, a great deal more experience and a great deal more evidence on which to build. Not indeed because we are smarter than our predecessors (generally we are not) but because we are more numerous. We have the legacy of previous generations, we have the work of the hands and minds of so many people over so many decades, to build on.

I have to celebrate that ours is indeed a Golden Age for astronomers and cosmologists. Again, not because of their ingenuity, which is high but not outstanding, but because of the tools we have been given to work with: extraordinary powers of electronics, radio, video, and satellites, that have given the astronomers an enormous strength, hardly foreseen by the ancients. The door was opened by Galileo still using light perceptible to the eye. Of course that's not the only light in the world. That is the light under which we evolved, sunlight, but we look beyond the sun into the depths of space and time. We see things in other lights, most of what I have to say is based upon the richness of evidence in the colors of the rainbow and then some! That is to say, from the radio, from the infrared, from the visible, and from the x-rays and beyond, the entire electromagnetic spectrum. Once in a while we go beyond, to neutrinos and, very rarely, gravitational waves.

I want to give a broad sketch of what I think a majority of cosmologists would agree with me is the present state of our art. I will leave out many things; I will not answer all the questions. We cannot answer all questions, or perfect every link of the chain. But I can give a broad sketch, persuasive in its symmetry.

### 1. *Our Current Act in the Cosmic Drama*

I have come to see that we live in a great cosmic drama, of which we know two acts. Our long act, the present one is the one we know best. The astronomers and the physicists have carried our insight out into space and back into time for ten or a dozen billion years. So we do know something about it. But of course, there are plenty of numbers left! To infinity, ten billion years is not much. We are constrained to dwell in a certain small segment of space and time, and try by the eye and the mind and sensory improvement by the instruments we have created for ourselves, to penetrate beyond the little volume of space time in which we have direct experience.

It is the galaxies and their clusters, and the voids between clusters that represent the great stage furnishings of our own act, the stage set. We have earth and sun and Galaxy, and many other galaxies and groupings besides.

This figure from a photo by David Malin shows a crowded field of stars. If you look with your binoculars at the Milky Way you'll see something very like that. Notice the black blot on the sky. When astronomy was yet young, William Herschel, a musician, the finest of all amateur astronomers, built the biggest telescopes in the world, to become the first professional with such a telescope. He thought those dark patches were holes in the sky, openings cut through the layer of stars to show darkness beyond. Of course we now know that's not the case. It's much easier to hide a thick layer of stars by patching it with a little ink than to drill a hole right through it that is aimed right at you! We don't have such self-centered feelings about the universe anymore. We see too many such holes, too many to be pointing at us. This is a dusty patch, a gassy, dusty region that obscures the stars. To show that such things can exist, in another part of the Milky Way, you might see a bright nebular gas illuminated by brilliant stars. Some stars are so hot we don't see them well by eye. They give UV, to induce fluorescence in the gas and thus produce lovely colors.

Notice in this photo by David Malin a shadow cast upon a mass of gas and dust by some brilliant illuminator, maybe a visible star, maybe one we don't see, blocked from view by an intervening cloud of dust. That's exactly what you'd see if you looked at the other dust cloud, not as we must look out at it, but if you could move magically to the side, to view from a 90° degree offset.

I put it here not because it's a specially important object, but because of one point I want to make. Besides myriads of stars, there is a lot of gas and dust. That's a very simple point to be told in a sentence of two. But I wanted to show one understandable example that requires a little thought about the shape of things in the world, what we call their geometry: not the theorems in geometry, but the actual arrangement. I do that to show how we apply the physics of everyday, of course cautiously and prudently, to the physics of this wider world. We find that more and more until now we can now look at things hotter than anything near at hand, or bigger, or denser, yet built upon simple extrapolation from the geometrical truths we have. I would hasten to explain that if I were to say this to my colleagues for the first time, they would be ready with all sorts of checks, to measure the faint light that comes out. Is it polarized? Are there stars we don't see? How many? How bright? and so on. They would pull the picture apart, making hundreds of their own, until they had satisfied themselves that all details of explanation fit pretty well. They will find a few discrepancies that lead to a modified model: there are two shadows, . . . But the principle would remain. When I cite evidence, the evidence is really deeper than the rather superficial show of a single brief synopsis.

Now we see the cosmic units, the galaxies. I remind you again that a whole galaxy is gigantic. A rich one contains ten or a hundred billion individual stars, aligned brightly along the arms, more dimly throughout the region, stars of different color different age.

We study those galaxies; each one might repay a lifetime of study, more than say the Antarctic continent by a vast magnitude! It's true that we know very little about it; for we cannot visit. We've studied maybe a thousand or two galaxies nearby. The whole Milky Way could be contained in a patch in the sky to be seen only upon magnification. These are nearby galaxies.

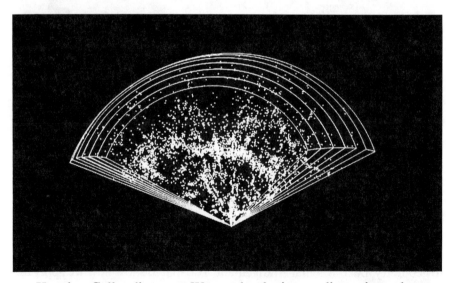

Here is a Geller diagram. We see the sky in two dimensions, the way we see the printing in a book. We think of a book as two-dimensional. We see a picture on it; and tend to think of the picture in 3-d, though of course it's not. We are clever enough, and the eye and the camera close enough, so we can make out a three-dimensional object in a flat picture. In the sky there is no such help. These are unfamiliar objects; we've never seen anything like them; faint patches. Maybe we detect some order in them as indeed we have—but we must first put in the third dimension. If you hold your hand in front of your face, you can hide the sun, but few people are of the opinion that their hand is the same size as the sun. Our two-dimensional retinal projection is not enough to understand the world. An astronomer can never understand the sun if he thinks it's the physical size of his hand! It's hard to know scale out there. The philosopher Anaxagoras was banished from old Athens because he taught (quite against the wisdom of his day) that the sun was a firey ball as big as the Peloponnesus! Probably he thought it was bigger still, but that was as much as he dared say.

Professor Geller and other optical astronomers in their wonderful galaxy searches have used a clever way to put in depth, the third dimension. They have drawn a slice, as a cross-section of the real three-dimensional scene. These galaxies are shown spread in space. Each point here is a galaxy, by no means all the galaxies we can see.

Now we see a broader view, not just a slice, but a tenth of the sky. Made at Oxford University, it presents about two million galaxies, their sky positions alone, no depth, no 3-d. You can't say which are close to you, which are far away. There are voids, there are clusters, not only in the two-dimensional projection, but in three dimensions too as the slice of space made clear. We begin to sense the universality of galaxies.

A deep picture reaches out so far that every spot is a galaxy. Though by eye we see a starry sky, we now look at the space between those stars to see a sky patch crowded with galaxies!

This whole scene is only a small sample of the sky, taken looking out through the thinnest part of our galaxy. A tiny piece of sky is much magnified by a big telescope, from only a hundredth of the area of sky filled by the moon.

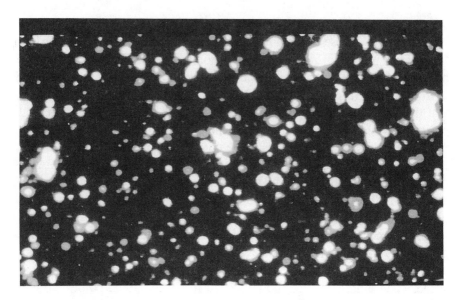

To see the "outside world" we naturally look the shortest way out of the
Milky Way disk of star in which we dwell. They build telescopes in the
wonderful thin air of the high Andes, to take beautiful pictures with modern
video techniques. The number of galaxies in the view, if translated into the
full sky, lets us infer that to this level, really quite faint, we could count about
ten thousand million galaxies, enough for a few galaxies per person. That
probably is not the end of the galaxies we can some day see.

I hope to have suggested to you one universal characteristic of this grand
domain: it is mottled, clumped, lumpy, discrete, or dotted. In the first view,
dots are individual stars. In the last, a whole galaxy of ten or a hundred
billion stars at once were still dots. We can't subdivide those; our instruments don't yet allow it. But the mark of clumpiness everywhere is sure. I
should say these dots are all made of matter, and they shine by light. We
know about light; we know about matter. An analysis of the light convinces
us that the matter we see far away is the same as the matter here under the
sun. All of it is built up in the most remarkable way, not as a a a molded
sculpture of clay, but as a masonry building, all built of distinct, commonplace,
modular, bricks. A hundred sorts of atomic bricks, all the elements of the
periodic table, plus photons and a few neutrinos and possibly other kinds of
"brick-like" particles make up everything that we see.

It is so out to the greatest depths of space and back to the greatest reaches
of past time, for looking out into space is also looking back into time. When

you face the mirror in the morning, you do not see yourself at your true age. You're a few nanoseconds older when you look than you are in the image. A nanosecond doesn't matter much, but the principle is unchanged. The calculations of the astronomer lead to the understanding that the further off you look, the earlier the images you see.

Now I'm prepared to describe the current act of the cosmic drama, half understood. We know much about it, though much we don't know. All kinds of stars, their birth and death, the elements they cook up and how, how the light flows out, what dust and gas do . . . I'm not going to touch on all astronomy. I want only to suggest what distinguishes this remote astronomical world. (I've not talked about the planets or the solar system nearby, nor the meteorites that plunge into the atmosphere of the earth.)

### 2. A Newtonian Act

I can say that behind all those motions and structures must be Newtonian gravitation. That universal attraction of every lump of matter for every other lump of matter is at the base of all we see. It is Newtonian gravity that first clumped the individual atoms of gas into the stars. The star in the end often spews its long-cooked atoms into space. Other stars will eventually form from them, as gravity reclumps the effluvia, to bring together, to network it as a new star. Our sun is of the third or fourth generation of the stars in our galaxy.

How do I know it's gravity that dominates? It is the only force that acts with importance among big distant objects. Now, why is that true? You know it is not true in every respect. Let me remind you of a familiar experiment. If you take a comb, to rub on your sweater. I hope you've then enjoyed watching the bits of paper near the comb, on a dry winter day in Minnesota (an ideal occasion for the discovery of electrostatics). In that dry air you will see pieces of paper fly up to the comb; often they stay there a little while, and then jump down again. If you've never done it, please try it one day. Try it anytime, but it will work better in the winter.

Let me discuss that experience as the physicist sees it. It's a remarkable demonstration of the physics of forces. We know today of only two kinds of forces that can act over large distances. (There are a couple of other forces that act only within the atomic nucleus: I shall pass over those.) At large distances, electric and magnetic forces, unified into electromagnetism, are one class, gravity is the other class. Those are the only forces that can be counted on somehow to attract the material that has collected into the lumps, clumps, and clusters—stars and galaxies— that you saw in the pictures.

But electromagnetism has a remarkable property not shared by gravitation. In the comb experiment the downward force of gravity is the result of a whole earth at work. We know that now. The force that fights gravity to pick up the pieces of paper is just what you get from rubbing two distinct small insulators together. The physicists know that only a tiny part of all the electric charges in the comb and wool are not paired, only a very small fraction indeed. Yet that small fraction is more than enough to overcome the pull of the entire earth. That really means that comparing atom for atom, particle for particle, electrical and magnetic forces are stronger by many, many, many orders of magnitudes than the force of gravitation. How then, can gravitation rule, as we claim it does, so that Newtonian mechanics and gravity can deal nicely with all we see in the dazzling collections of dots in our cosmos. (Of course within each dot there are other forces at work. The earth is kept from collapse by atomic forces within. Matter doesn't crush completely, but it resists gravity.) Nevertheless it's gravity that assembled the thing, until a balance has been struck within each dot.

The reason is easy to understand. It is clear in the most famous of catch words of elementary science: Like charges repel, and unlike charges attract. (I hope many would have learned that.) You never heard that for Newtonian gravitation. Newtonian gravitation says instead that every particle of matter attracts every other particle of matter. Compare the two situations as a system grows. Large systems have grown from something smaller. It's unlikely that they were cut apart from something even larger; there's no room for that! Suppose a positive charge attracts a negative charge. Good. They will fly together if they can, much better than the same number of uncharged particles would if they were only gravitationally attracted. But once it gains the negative charge, what is then the charge of the system? It is reduced, because the initial positive charge is now partly canceled by the negative charge. From the outside the net charge is dwindled. That will keep on, until finally the collection of unlike charges has made a mix that is nearly neutral. That's the way the earth is, the way your hand is, almost every large object in the world, not atomic and microscopic, is neutralized. Were it not neutralized, if it had a surplus either of negative or of positive charges, it would attract to itself the opposite kind and soon become more nearly balanced. We say it is easily saturated.

But gravitation is utterly different. It is *insatiable*. You can't saturate it. The heavier an object is, the more it draws all other objects. The more objects that are drawn in, the heavier it becomes. The heavier it becomes the more objects it attracts. The situation is clear. This is a runaway feedback system, guaranteed to amplify clumps of matter. Something has to intervene to stop it. That could be some other kind of force, more likely namely the exhaustion of the supply of matter nearby, or swift motions of matter directed <u>away</u> from the attracting center. That's enough to show that Newtonian gravitation is a sufficient explanation for the things that we see. It is also the main force in forming all those more distant galaxies and clusters that we do not see yet. We know of no other force grand enough to carry out the same task. I think I say with justice that we live in the Newtonian act of the cosmic drama.

I would add one check on this. If you consult the astronomy books, you'll see much detail to bear it out. I would like though to show one simple result, of very wide occurrence. Produce a grand conclusion,—Newtonian Cosmos—it makes everyone happy, and I believe it to be true. You can't stand on that statement alone. You must confront that statement by a test. Is there some way of testing my obvious proposition that insatiable gravitation means that uniform stuff will coagulate into lots and lots of clumps, which enables me to name our Act after Newton.

You all know this test, a very clever one. If I draw moving material into a center, some of it comes in tangentially, in a glancing way. That brings with it a certain amount of what we call momentum of rotation. Like energy that is conserved, never lost. You've seen it in the beautiful skills of the ballerina or the high-board diver, who have learned how by bringing in the limbs they can spin faster. When she spins too rapidly, she can slow down, merely by extending her limbs. Converting energy by moving matter apart, she slows it down. Bringing mass together *speeds up* the spin of the ballerina or the moon, whatever was spinning.

We know the earth spins. We see night and day. Of course the moon spins, the sun spins, the comet Haley spins, the asteroids spin, the stars spin, and galaxies spin, and we're pretty sure the clusters of galaxies spin as well. We know all that from observation. How come that everything is spinning, almost nothing is at rest? A few items that are at rest are the exceptions that prove the rule. You can show in almost every case there is some process drawing away the spin. Broadly speaking, in our cosmos everything spins!

If the cosmos really were made from a much more diffuse material that had been collected into lumps, would those lumps not spin? They would, unless they had come in with perfect symmetry. There's no reason why all the motions of all that came in were all exactly toward the center. If they were unbalanced in the least one way or the other, the resultant spin would show up in the amplified spin of the compact center, just as the ballerina's beautiful pirouette emerges by drawing herself in and up on points. She has reduced the average radius of her body. The uniform result of astronomy, everything spins, is then no mystery. It has to spin if it was accumulated from materials not moving in perfect symmetry. Moreover, the spin we see is not uniform. It is oriented every which way, the axes of spin point in all directions. What is gained by one object is lost to another; the total is preserved, but there's no tendency to one axis. (Groups like the solar system alone have common spin.)

We are seeing an act of the cosmic drama due to Newton. The order of the day is that some kind of well-distributed matter was clumped by gravitation, with all the motions, chemical changes, pressures and shocks and explosions, and everything else you might add during the course of this operation. We're not talking about something in a soup bowl, but something in the grand theatre of the whole universe we see. I'm prepared to say "yes, that looks like the act we're in."

Something quite interesting is in the motions we see besides the spin. There's a slow drift apart of the largest elements of the cosmic furniture. The groups of galaxies drift apart, one from the other. Stars don't and planets don't drift away from their neighbor suns, but the galaxies drift apart by expansion of space and by initial motions, as Professor Harrison described. That is certainly what we're seeing, a rough order to this drift, heavily used by the cosmologists. It's a discovery of the 20th century. Only since the 1920s have we known that this was a fundamental way in which galaxies moved. They spin on themselves; they orbit very majestically and ponderously around each other in a group; but then the distant groups slowly separate in all directions, as the whole collection becomes dilute. That's the state of the act we are now in.

### 3. *Before We Came In*

The cosmologist, the physicist, is always trying to extend knowledge, ask more questions. What came before this act of Newton? Was there anything? Is this the single long act of the universal drama? Maybe. Let's ask what happens. Run the tape in reverse, if you can imagine doing that. See the expanding matter come closer and closer together in reversal. If you run it all backwards, you would long ago find it close by. As it becomes closer and closer, the pieces merge. They overlap each other. As they merge, the gas is denser. There are more collisions, much more is going on. That's evident. It seems quite obvious. It would be very nice to find evidence for that! That is the wonderful thing that has happened during my time as a student of these matters. It began with strong hints in the mid-1960s; by now it has been beautifully demonstrated with quantitative accuracy. We discovered a distant region of our world emitting a tremendous amount of radiation! Much more radiation, a hundred-fold more radiation than for all the stars in all the galaxies. It is coming from a distant place hence a long past time. It lies in all directions, wherever we look—by summer, by winter, from North America, from South America, no matter which sky direction you look, you'll see the same thing. That uniform result is not a small detail, it is difficult to explain. No, no; it is the major source of photons in the universe! We can't neglect it. As we looked, we saw more and more, most completely in 1990 and '91 that it was an extremely bland, uniform, simple and spinless region of matter. It is absolutely the opposite to all the matter we've seen till now. What we know was chemically complex, not uniform but lumpy, structured, grouped or showing voids, spinning. This particular stuff, the most important emitter of radiated photons. (To be sure, none of the photons are visible; most of them are in millimeter band, where only recently are we able to measure quite well.) That millimeter band radiation is the so-called background radiation; it has given us the strongest evidence have. It is the sign of a previous Act of the drama. That prior Act was not dominated by gravitational attraction, or by clumping that made spin and structure and diversity everywhere. That older Act was marked by uniformity, blandness, the absence of spin, the absence of detail. This was long past behind all the galaxies. The gas itself we see is not a character on that ancient act. The curtain has come down on the early act. But the early act prepared our material, for this must be the primal gas from which all the stars and the galaxies were later made by the action of gravitation and the chemistry that went on within the gas. That would explain everything, and the details fit quite well.

We have seen at least the footprint, if not the actual material, the footprint of an earlier stuff that once ruled the entire cosmic drama before our Act. It is Newton's Act that made suns and stars, planets and life, elements like oxygen and carbon. None of those were present in that earlier, simpler universe.

How could that be? Can we explain it? It turns out there was already in the textbooks a beautiful way of explaining it, not recognized by us until pioneer work by several astrophysicists in the Soviet Union in the late '70s. The capstone of real recognition was placed by Allen Guth of MIT, who put together the clues and showed us how powerful the argument was. Using the equations written down by Einstein in 1917-1918, adding one very great idea and strange idea from contemporary particle physics, he saw that those equations admitted and indeed demanded an amazing answer to the question, what came before our act? We are not in Act I. I hasten to say we are not Act II either as far as I know. All I know is there was at least one act before us. Maybe there was an act before that too; I can't tell you that at the moment.

But I can tell you something about this previous act. How would you make something clumped, diverse, and spinning? We learned that: by attraction, a pull that bring things together unstably, never saturated. How would you make something so much the opposite in appearance? Why, the answer is very simple.

Do exactly the opposite thing. Fling everything apart; very suddenly stretch it all out. Suppose you consider the Himalayan Mountains, a very complex relief map indeed—Mount Everest, K-2, many valleys in between—terrifying. But imagine one simple thing; stretch the Himalayans from the real 1,000 miles across the vast range; leave them high, but stretch them out horizontally, 1,000 miles, 2,000 miles, 5,000 miles, 1 million miles, even much larger than the whole solar system. What would happen to the relief map of the Himalayas? It would now look very bland. Of course, Everest would still be 5 miles higher than the next valley nearby, but "nearby" would now be a million miles away! The landscape would look extremely flat, hardly changing utterly bland. That's exactly what must have happened to take a very large part of the universe, which was surely not so beautifully arranged, and suddenly make it bland and uniform, and moreover, bring it all to exactly the same temperature, to about three decimal places, at all parts this gas!

Is there an answer to that? I have singled out the contrast, to lead to you to think of it. The answer is: all you need do is make a great change in your mind, and say that in those days whatever was matter did not attract gravitationally, but repelled: each particle repelled each other particle. What is the result of that? Of course, a tremendous sudden stretching. The more you stretch, the faster you stretch: that was exactly in the equations of Einstein as a possibility as soon as you put in the condition for repelling matter and not attracting.

So that's what we believe happened. This is named by an obviously timely word, inflation. But it really represents repulsive gravitation. That repulsive gravitation smoothed and inflated some tiny volume at an earlier period, to turned it into a bland material which they featurelessly decayed, into all the matter we now find, through a whole series of stages, out to the remote distances. It left us with a beautifully uniform region. All we see was once in easy interaction, a very small region of space that expanded hugely. Maybe it was no larger than a single proton, or even smaller than that. It expanded to giant distances, containing all we see, and all that we shall see for some time to come. That was the previous act, the act of anti-Newton. So our act is Newton's and that earlier act is anti-Newton.

Suppose you were walking along some trout stream in Northern Minnesota and you saw among the many pebbles a most beautiful pebble of jade, much larger than any other pebble. You would certainly be amazed at this object. You would take it home to scrub and polish, to see what it was; that's about what we did once we found the background radiation!

Here's what NASA telescopes in orbit see. This is an infrared map of the sky; this is not the great result, but only to explain the result. Here you see a line across the map that is the plane of Milky Way. We are seeing here a projection of the Milky Way. Across the center of the disk runs the plane of the Milky Way.

What then is the blue line? That wonderful curve? It is nothing very strange. The shape is not given by nature; the shape is given it by the way we have to plot our map. This simply represents the fact that besides the Milky Way, where the infrared comes from distant stars, it comes also from the meteorites, cometary dust and debris that lives in the solar system. That debris lies in the plane of the planets. In that plane then a lot of infrared is plotted here in blue.

I will now show you the same kind of map taken in the millimeter band by the Cosmic Background Explorer satellite at the end of 1990, over many patient traverses of the sky. You'll see what happens when we look with a rather blurry view at patches of sky bigger than the moon. Thousands of patches are put together to make the map that you will see.

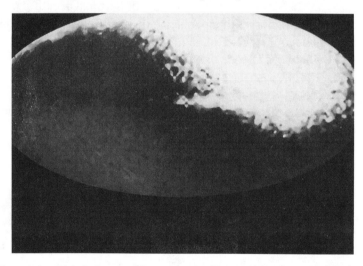

There it is. This is hotter where blue, where red cooler. Notice the shape of this boundary: rather reminiscent. One little extension here is clearly not part of the general form. That is the center of the Milky Way, where there is a local source of this kind of radiation. It has nothing to do with the great cosmological issues; so we subtract it off as interference. Another station broadcasting in our wavelength, so we'll subtract it. But now what about this boundary curve? I hope I've shown enough to convince you that the shape is no mystery to us. That shape simply means this great sphere which is the jade boulder we found in the stream. The two different hemispheres are parted by a simple equator between them, just a plane that cuts the sphere in two, like cutting an apple right down the middle, not necessarily along an axis, just down the middle in any direction. You get this shape in the usual mapping.

What could there be to turn the great distant cosmic background into a wonderful hemispheric division—one hemisphere warmer, one hemisphere cooler than the rest. We know. If you move toward a source of radiation, it becomes warmer to your view. If you move away from a source of radiation, it becomes redder, hence cooler, to your view. We are surely not standing still. We're moving in the galaxy, around the sun, and more. This division is the sign of our own motion, projected on the sky. But that motion is not very great, one part in 600, each way, and that's all. So that the ball you took home from the creek was a perfect sphere, except that it had a little coat of paint on half of it. When I rub off the paint, I expose a beautiful white jade boulder, perfectly round to a part in 100,000. It's a wonder of the world, the greatest wonder of our world, I think. If I subtract this motion effect from the real sky, what I have left is a perfectly bland map, with no features at all, more uniform than plain white paper! The source is rounder, so to speak, than any ball bearing you can buy. It's no small part of the universe either; it's all the gas we came from, and it produces 98 percent of all the photons that we see.

You could say, of course, that this is simply the way the world was built; we'll never understand it. It's not given to us to understand why we come from a completely uniform gas, to start clumping in the Newtonian way that formed our cosmos, and gave rise to ourselves. But the physicist is trying hard not to leave anything to such arbitrary flattery; we say this was not made simply so we'd have simple calculations. There was some reason for it. This was not under any external edict except for the laws of nature. Nor was there a meaningless combination of accidents. It was a physical necessity of that time, forced by the nature of the forces that existed at that time, just as the fall of every apple is forced today. This is evidence of a previous act, the great jade boulder which is the cosmic background was left over from the previous act. Before we could come into being that gas (which we have good reason to believe is nothing but a mixture of hydrogen and helium at about 5,000 degrees Kelvin) glowing white hot. We see it as from so long ago, moving so fast away fro us, that its radiation has become millimeter-band radiation. We now believe we know one act before the distinct act that we are in now, an act so far 10 billion years long, and still rolling.

## 4. *Enforced Simplicity, Inaccessible Origins*

Now we walk onstage, to examine the world we inhabit, all built by Newtonian attraction: all those glowing, spinning centers of clumped matter. The farthest portion of it that we can now see is the uniformly glowing, expanding and cooling gas that emits that distant background, the legacy of the inflation. (Recall, we cannot see anything of that anti-Newtonian Act that preceded our own. Probably that sort of matter doesn't even generate electromagnetic waves.)

But the physicists didn't surrender. Guided by Einstein's theory and the rich knowledge we have of radiation and ordinary matter, they looked back by calculation alone, not to any previous act, but to earlier stages of our own Newtonian act, that denser, still expanding, gas before the opaque layer we see. They could fit a number of features of the chemical recipe of the cosmos quite wonderfully. We have samples of matter—some of it is in the electrons and protons that run in your own veins—from before the background glow, though there is no radiation from that early time that we can now detect. Of course the result was entirely based on extrapolation back to the denser, hotter state. But the results seem to check, and indeed the ingredients of the physics were what we richly know.

So the experts extended their successful extrapolation back, back, back—it now seems to me quite shamelessly—over many, many powers of ten in time, if only by a small fraction of a second back, back to the BIG BANG, an all but zero time. There it was agreed the extrapolation had to stop. For before that near-zero time, by the nature of the Einstein equations, indeed, by their failure, there was no guidance from known physics, no time, no space, no matter, only a boundary against the grand unknown. That was what we called THE BIG BANG (the term was first coined rather derisively by Sir Fred Hoyle.)

Though the metaphysical qualities of the Big Bang give it a universal interest, an overconfident extrapolation was its only support. Now we see the deepest past all very differently. Instead of a smooth extrapolation back as long as the equations work, we come instead to an early and catastrophic change in matter. In that early universe there was not simply strange hot matter, but something specifically new, an anti-Newtonian matter (sometimes called the false vacuum) that drove the explosive inflation to an enormous scale, soon to decay everywhere into more "normal" hot gravitating matter that expands still with far less vigor, merely coasting against the drawing-in by gravity that smoothly slows the expansion down but is too feeble to stop it for a very long time, if ever.

We are in the position of experimenters watching the film of a kettle of water cooling off, but running the film backwards. They see a steady warming, warming . . . and they extrapolate to an incredibly hot beginning, all by continuous change. What they had neglected—for they did not know, either from experience or from theory—is that heating water hot enough turns it to steam! No smooth extrapolation can bridge that change.

Just such a change—far stranger still—was inflation. No longer can we say we know what came before. The evidence is lost. We expect that in that very first moment before inflation, the cosmos was fully within the laws of space, time and Einstein's gravity. For the inflation is physical enough, and governed by familiar laws. It is no mysterious boundary, but a single physical event as much localized in space and time as the fall of an apple.

Even their long extrapolation was not fully wrong; we still believe most of it. What we *don't* believe is the *final* extrapolation. We inserted inflation instead, and now we must ask how inflation began. Inflation, I should hasten to say, is not a proved scenario. It is, I believe, a plausible and perhaps even a probable one. We have to prove it out in the next decade or so. Inflation has given its hostages to fortune, predictions of how matter is distributed out there. We will learn that one day, and then we'll know. Perhaps inflation will go away, back to the old story. Perhaps something new will appear. More likely, I think the future will bear out inflation.

Then we would have to realize that the repulsive phase that made inflation, stretched out the Himalayas, flattening any arbitrary disorder into the beautiful blandness we see, was nothing at all universal. It was a big bang all right, big enough, thank you, and our own, but in some grand sense it was still local. That was not The Big Bang it applied only to the small part of the universe that we know and see now; it was not at all universal. It was not at all a boundary to space time; it was not at all the beginning to everything. Before it was a universe, not exactly ordinary, but built of plausible if new particles and fields and temperature and time and space a physical cosmos. At one point it acquired a prodigious ability to expand. That expansion dominates all we now see and all we ever will see for a very, very long time.

Now the big question is open. What was the universe *before* that state that led to the inflation? We don't know. Was it a steady state universe? Maybe. Was it expanding? Maybe. Is it closed? Maybe. Is it open? We don't know the origins, if any. Was The Big Bang an earlier origin of all? We don't know.

Our cosmos has a preferred and symmetrical frame, a Euclidean geometry, to a very high degree. We live in a spacious pool of galaxies, finite but much larger than we ever thought before. We have a vast future of time, finite but much longer than we had ever thought of. Some say 10 to the 10th years! (A googalplex, I think it's called.) We don't know. But we do know it is a giant affair. Is there something more universal, grander than that? We don't know. Was there ever any opening to the play? Maybe there were inflations before our inflation. Maybe there are many acts. One is not the second act but the Nth? I don't know. Maybe, but maybe not. We do not now know, and all the clues for now are hidden behind that uniform and enigmatic blandness, the decisive simplicity of the inflationary act.

That's how it stands, to justify the second part of my over fancy title. Now that I have explained, you can see why I used those words. We live now in a universe, a cosmos, of enforced simplicity. It was enforced by inflation. Inflation was driven, not by any accident, not by any external tamperings, but by laws of physics that demanded a uniform expansion, just as now the apple falls. We live in a simple universe we always felt was simpler than it might have been, simpler because of this early action:   an enforced simplicity.

At the same time this simplicity conceals the detail we'd like to see before it came. We may be able to penetrate that barrier, by new means. We've done great things like that before! But at the moment I would say that the origins are inaccessible. What I see are two acts. We don't know how many more:   maybe only those two. We live in a long act controlled by Newton, after a preceding very swift Act controlled by anti-Newton, and that simple symmetrical narrative is all we now know of the cosmic drama.

# Questions and Comments

**Fuller:** First we will entertain questions from the panel.

**Ferris:** What I have is not really a question, but a comment in that I think we've seen a demonstration of why Philip Morrison is known as a philosopher as well as a scientist. In listening to his talk, I was struck first by the fact that I had not myself appreciated the importance of the inflationary hypothesis, there's a technical term for that, about what he was discussing. But also for the fact that Allen Guth, who as you heard, came up with the idea, did not himself appreciate its importance. It's an interesting fact to me about scientific discovery that it really is discovery in the sense that often the person who has made the discovery does not realize its significance at the time. It takes a while for big ideas to sink in even on the part of the person who made the discovery. So one of the things I enjoyed about hearing this talk was beginning to realize just how important inflation may be.

**Morrison:** May I make a comment on that? Allen Guth has the office but one from mine; I see him a lot and we're always arguing. Allen is a believer; I am more of a skeptic. Allen has good reason to believe; he's done wonders in the complete panoply, GUTS and the plateau and the standard model and the Planck length and all those things, just as written in the book. Therefore, it certainly was a Big Bang, it certainly went just the way the book always said, and then it inflated! I say maybe. There is no evidence for that, but that's what we believe, because we believe field theory and know of no way to make a proton decay slow enough, or whatever, without something like that. He will admit, if pressed, that the simple extrapolation is gone. You have to say that before the inflation there was a hot universe full of scalar particles and Higgs bosons, because that's the way it should be. Maybe he's right. I'm not saying there is no Big Bang. But I am saying that all the evidential nature of that Bang has disappeared.

**McMullin:** I have two questions for Phil Morrison after that very exciting talk. One comes from the earlier part and I think it's a question that probably has been occurring to at least some of the people here. In the first part of your talk you mentioned that Newtonian gravitation is entirely responsible for the structures and motions of what you call Act I. By the way, I'd rather call that Act II just to get things in a better temporal order.

**Morrison:** But it's not Act II! How could we know?

**McMullin:** I know, I know, all right. N + 1. All right. I'll settle for N + 1. There have been a number of challenges to the thought that Newtonian gravity is in fact responsible. One rather sensational title recently of a book claiming that the Big Bang never happened in part depends on the claim that other forces, in fact, the very force you were talking about, electromagnetic forces are responsible. How likely, or how would you comment on alternatives of that kind which would attribute the cosmic expansion, the microwave background to factors other than galactic expansion?

**Morrison:** Well, I don't have the predisposition to deny them all flatly, but none have been shown to have predictive power. I don't think the plasma can tell you anything good about the microwave background. It won't tell you the depth; it won't tell you what it's made out of. I don't think that anybody in the trade except its few progenitors feel it has much hope. I think gravitation rules there, maybe with some radiation too. Gravity might have to be modified. There's a small sign, "a cloud no bigger than a man's hands." But most everyone thinks the modification will not come by changing the laws of gravity, but by adding more matter than we can now see. That's the big test that, I feel, and most proinflation people feel: there's much matter out there that gravitates, for we can see its effect on motion, yet it doesn't shine! Is there such matter? Well, if the particle physicists are right, there's all kinds of that sort of matter. They enumerate them almost by the dozen, and you should expect to see some. We know we can't see everything that's out there. There must be matter that doesn't shine, because it's inconvenient, in scale or history. We know there are familiar particles like neutrinos that could be out there in great plenty (though technically, they fail to fit.) But something like neutrinos from an earlier condition would explain everybody's feeling that gravitation from what we see is not enough. There's more matter, strange stuff; maybe five or ten times more than all we see. This is not to state a dogma; this is to be tested, soon we've got to find that matter in some way or we won't be satisfied: by A.D. 2000, with luck!

McMullin: The other quick question I have—I caught a certain tension in the way in which you described your assurance about Act II. You used words like pretty much, for sure, or the laws of physics demand, talking about inflation, but yet on the other hand, you said it was not proved probable plausible. I wonder if it isn't too strong to say that it's demanded by the laws of physics. That really does seem to be a little too strong.

Morrison: I think it is too strong. I agree with you. It's allowed by the laws of physics, implied by certain formulations of these laws like the standard Higgs-boson model to generate mass, but it isn't really demanded. You're right. I mixed in too many metaphors for one short talk, but there is that jade boulder! It is most extraordinary. The data on isotropy, the data on temperature, are far, far more quantitatively precise than any other data we have in cosmology: better than the red shift, better than the density of matter, better than the distribution of the galaxies, better than the correlation between galaxies. Those are always statistically based on so much less information. No theory that does not say why the background is uniform to a part in 100,000, or even parts per million is going to stand. It has to do that at least, then it has to be consistent with everything else as well. But that brute fact is enough to tell us that something like inflationary stretching took place. Why it took place is open to more questions. I can look in the books, or talk to Allen Guth. He'll tell me about the Higgs-boson, but that's not the only answer.

Geller: Phil, I wonder if you would comment on one of the predictions of inflation, in which inflation makes a prediction of how much matter we should observe per unit volume in the universe and those of us who are professional I think, well, all scientists are professional skeptics I guess in some sense. Perhaps observers are often more skeptical and one of the problems is that we don't detect this density of matter that's predicted by inflation. Now of course we can't show that it isn't there, but it's a rather odd logical dilemma that we're in and I wonder if you could comment on your view of that. I think that that's the thing that raises a lot of doubts about the inflationary epic and people who are actually actively observing the universe.

**Morrison:** If you have doubts about the inflationary picture, you should have much more doubt about the standard Big Bang preinflation, because it gives no reason for the utter isotropy of the black body radiation. It says the background should be 10 or 20 percent from uniform the way the galaxies and the clusters are. There is a contradiction to that by 4 or 5 powers of 10, enough to make one feel there must be a deep reason for that simplicity. Now we give you a reason. The reason entails other things that are hard to observe, we grant that. I'm not trying to say you should give up and conclude not to observe! No, no, this is a crucial test maybe some great new ideas will come out of the feeling that there is isotropy but no extra matter. That would be very valuable. But don't forget it's already rather troubling that the universe is so near to flat. Forget inflation, forget all those things. Just take two numbers, the Hubble constant and the q value. It was all we said, we order the magnitude of being flat anyhow. It was always crudely flat for q, was never far from zero. That was already quite a problem. I won't tell you about the horizon and all those more technical things which make me feel something genuinely special about the black-body radiation. The Planck distribution is satisfied by the cosmic black body radiation better than by any laboratory experiment. That has to be explained. The explanation is ready at hand, given by the Einstein equation, the same equation we use for everything else. The only thing we don't know—and I agree we don't know—we have only conjecture—is that it was caused by a scaler field. But you don't need to say that. You can just go ahead, fiddle with all the forms of the Einstein equation and inflation will come right out at you or at least at Allen Guth.

**Harrison:** I agree that our present picture of the universe is truly amazing as you have so ably explained and yet I'm also aware of the fact that the cosmological devines of the past have always been amazed at man's knowledge of the universe. And have always looked back on their ancestors and pitied the pathetic shreds of their knowledge. Here we are doing much the same. My question is this, is Professor Morrison, in his certainty of current knowledge, denying our descendents the right to look back and pity our pathetic view of the universe?

**Morrison:** Of course you're absolutely right. I wouldn't deny that for a moment. I began by saying you had shaken me with this long muster of very clever people who said what we regard as absurdities; now here we are, probably not as clever, saying even more absurd things. In spite of it all, I think there is a cumulative nature to our physical knowledge about the universe; they did not know about the dominance of black body radiation!

**Harrison:** But all our theories may turn out to be wrong. We have no evidence that general relativity is the proper theory of the universe.

**Morrison:** Don't really need GR!

**Harrison:** It works locally; we've proved it locally in the solar system, but not for the universe.

**Morrison:** I agree with that. But if I wanted to make, say, a kinematic relativity, I would still come out with something like inflation, some way of stretching gradients to get rid of them. That's a natural idea, and that's what I'm trading upon finally. I go a little further with the defects of my time. I believe in GR to a degree, but not yet to black holes, nor to singularities. I would not carry the narrative back down to the Planck length at all. Your question is really, is there room for us to work? There are satellites to Jupiter; I think you'll agree with that, even though the philosophers demonstrated it could not be true. There is some change in our views, not as much as we think, and I would admit they must be tested. Maybe in the past, they took the arguments of powerful authority too strongly. They didn't test them as much as we now do. Nowadays you can't get away with that: there's always somebody like Margaret and her friends, who is going to demand that theorists put up or shut up. That will save the situation.

**Fuller:** We have several questions from the audience that would like you to talk about the transition from the anti-Newtonian universe to the Newtonian universe, or is that part of the movie we slept through?

**Morrison:** That comes before we ever see anything! We don't see that reel of the movie at all. What we sleep through is between red shifts we see of up about 5, beyond them to red shifts of about 1,000, between distant quasars and the black body radiation itself. We don't see a darn thing yet in there. We have to look for such things. We'll never understand galaxy formation, or void formation, till we fill that region with some kind of information. What, I don't know. But that's to be found out. We don't know what's going to come next. It's not a priori. I don't know what's going to be there. I agree to that. But that is not the inflationary time. We were not even awake yet during inflation; we had not come to the movies. As far as what happens there, all you can do is write down a simple exponential expansion. But to say we know just what happened is to go too far. We have reason to believe that a scalar vacuum field would turn into the ordinary energies we have now, given only a very small fraction of a second. We don't really know all that happened, but the consequences are so clear and so valuable that we'd like to think we know what might have happened.

**Fuller:** One more question. How do you know that the background radiation is spinless?

**Morrison:** Because if the microwaves were from spinning gas, it would have an initial motion on top of its expansion, a pole where also there was no motion due to spin. There's no sign of that. This gas is so far away, and yet has no ordered motion across our line of sight, and that gives it a very small spin. I've forgotten the limit; it could turn once in 10 to the 18th years or something less. That's a pretty low spin!

# EVOLUTION OF INTERSTELLAR COMMUNICATIONS SYSTEMS

## TIMOTHY FERRIS

The talk I'm going to give today begins and ends with the subject of information. During the lunch break today I was sitting outside in the sun and a ladybug landed on my copy of this speech. I stopped reviewing the speech and examined the ladybug. It's a beautiful piece of design, a ladybug. It makes a Porsche look crude by comparison—and I'm speaking as someone whose idea of pure joy is to run a Porsche around the twisting turns of the Sears Point Raceway.

All the information it takes to make a living thing as wonderful as a ladybug or a human being is contained in the genetic sequence in that creature's DNA. As you may know, there is a project now to "sequence," as it's called, the human genome—that is, to determine all information that makes up the DNA of a human being. To sequence the human genome will take many years and a great deal of effort, but ultimately one will have a long string of digits that can be put on a computer disk and that will amount to the recipe for making a human being. At that point we will have translated that genetic information into a series of zeros and ones that a computer can read.

We live in an information age and I'm starting to get the sense that something very important is happening in our epoch, something that has to do with information and that will have wide-reaching consequences in science and elsewhere—in much the same way as technology and science have interacted in the past, in the age, say, of Galileo or of Archimedes.

So, what I'm going to talk with you about today will end up being about information. However, my subject is not a particularly scientific one. Instead this is to be a speculative talk, concerning what is called the search for extraterrestrial intelligence, or SETI, which Phil Morrison described some years ago as being more akin to exploration than to science. SETI certainly is an explorational sort of endeavor.

SETI was originally proposed in 1959, by Philip Morrison and Giuseppe Cocconi. I'm going to proceed on the assumption that there is something to SETI. That is to say, I'm going to assume that there are intelligent beings elsewhere in the universe and that it's possible to communicate with some of them by radio technology. And I'm going to make a speculative argument based on that assumption.

The subject of extraterrestrial intelligence has become part of our culture—as evidenced, say, the *New Yorker* cartoon showing an alien on another planet receiving an episode of the TV sitcom "Leave It to Beaver." This cartoon makes the point that we humans have been broadcasting inadvertently into space for something like 80 years now. Thereby we have generated a sphere of radio and television signals with an 80 light year radius. Within that sphere it is theoretically possible that an alien civilization could receive the weak, inadvertent emanations coming from our radio and television transmitters, our military radar installations and so on. In the cartoon the message from Earth reads, "Gee, Wally, don't tell dad I blew up the lawn mower, okay?" So we have in some sense announced our presence, though as yet we have no way of knowing if there's anyone listening.

Since 1959 a few thousand hours of radiotelescope time has been devoted to listening for signals from space. Nobody has heard anything yet, nor is there any other evidence to suggest that there is life, much less intelligent life, anywhere beyond Earth. So everything I'm talking about is speculation.

What, then, makes us think there might be life elsewhere? Mainly the numbers. There are just so many stars out there—several hundred billion in our galaxy alone—that even if life is rare, we would still expect that there would be a large number of life-bearing planets. The same chemical elements make up these other stars and planets, and they function according to the same physical laws that operate here on Earth. So one can imagine chemistry like we have here on Earth taking place in other places. Life itself is a fancy form of chemistry, so we assume that we may not be the only planet where life has arisen.

We are further encouraged in this hypothesis by what might be called the historical continuity between inorganic and biological events. At no point in the history of the Earth and the solar system can you insert a knife blade and say that here something miraculous happened—that prior to this moment everything was natural and then a miraculous spark of life appeared. Instead, it appears that the Earth gave birth to living organisms naturally, and quite early in its history.

But these same numbers cut both ways: Billions of stars means we must expect to listen to many stars before we happen across an inhabited and communicative star system. Even if there were, say, 10,000 civilizations in the Milky Way galaxy, all beaming signals our way that we could detect with existing technology, we might have to listen steadily for a generation or so before we happened to aim our antenna at a star, a planet, that's sending us a signal. So SETI is likely to be a long-term affair. It's not the sort of observing project where we can justifiably throw up our hands and give up if we don't get quick results.

If we do receive a signal, where might it be coming from? According to the customary SETI scenario, the signal would originate from an inhabited planet. I call this the "Lonelyhearts" scenario. It assumes that there is a species out there so lonely that they've long been sending out a message, waiting for somebody to respond. Their message reads something like, "Lonesome, technically proficient species seeks same. Object communication."

Now this seems like a reasonable scenario at first blush, but it has some problems.

One problem with the Lonelyhearts scenario has to do with paranoia about broadcasting. Although we have as I mentioned inadvertently sent radio transmissions into space, we have done very little _intentional_ sending of signals. The main reason that there have been so few such attempts is that we fear the unknown. That fear is not necessarily unjustified, in my view. After all, we don't know what's out there. Do we really want to start saying, "Here we are; here we are; over here!" before we get some idea who or what is listening? Probably not. So SETI programs listen; they don't transmit.

Secondly, there is the question of expense. If you want to send a signal omnidirectionally—meaning that you're broadcasting in all directions at once—at a power sufficient that a species no more technically proficient than ours would have a decent chance of hearing it, that takes a lot of power. So we're asking another civilization to invest a prodigious amount of energy in broadcasting.

Finally there is the problem of long Q&A times. The more optimistic SETI scenarios, the ones that anticipate that there are something like 10,000 communicative civilizations right now in the Milky Way galaxy alone, still end up saying that the distance to the very nearest one is something like 500 light years. That means that if you get a message and you send a reply instantly, it will have taken 500 years for the message to have reached you, and your reply will require another 500 years to reach the sender, and so forth. So SETI is not like carrying on a telephone conversation. It's more like reading the works of the ancient Greeks.

What might a message from space contain? In the classic SETI scenario, scientists tend to envision the message as having a primarily scientific content. That makes sense, but I wonder how much of a market there's going to be for purely scientific messages. I mean, if you could choose to receive one message from someone who lived a thousand years ago—Omar Khayyam, let's say, who lived in the twelfth century and was known both as a mathematician and as a poet—would you prefer that Omar sent you the general cubic equation of the third degree? Or would you prefer that he'd dispatched a verse of his poetry? Like this one, for instance:

> Ah love! Could thou and I with fate conspire
> To grasp this sorry scheme of things entire
> Would not we shatter it to bits--and then
> Remold it nearer to the heart's desire.

My point is that we already know the general cubic equation of the third degree. We don't need Omar for that. We could have done it without him, and indeed, several mathematicians did; the equation was rediscovered a couple of times, and only recently has it been appreciated that Omar did it first. Yet who else could have composed those lines of poetry? Art, you see, is unique and individual in a sense that science seldom is.

So I expect that the kind of traffic we would eventually find passing among stars would be cultural as well as scientific, if only because art ages better than science. Omar's is still a good poem. His solution to the cubic equation is still good too, but it's kind of irrelevant now, since we already have it in hand. Art in its uniqueness rescues each culture from the strictures of mere progress.

The longer we ponder the prospect of interstellar communication, the more we realize that the great gulf likely to separate communicative civilizations in the galaxy is not space but time. I'm thinking of the question of how long such civilizations typically last. We here on Earth have been on the air for only about 100 years. If we blow up the world or stumble into some kind of terrible problem or another within the next century, then our example will indicate that the average on-the-air lifetime of a technological civilization is only about a century or so. If that's typical, then there's nobody else in the Milky Way right now. In that scenario, intelligent life is a sometime thing: Civilizations scintillate into existence for a very brief time before disappearing. If technologically developed civilizations typically last longer, then there are more of them now. If, for instance, the average lifetime of a civilization is 10 million years—if in other words they're so on top of things that they normally stay on the air for 10 million years—then there may be thousands of civilizations in the Milky Way galaxy communicating today.

Yet even if we entertain such an optimistic value, and thus assume that there are thousands of inhabited, technologically adept species out there with an average on-the-air lifetime amounting to millions of years, we still come up against the curious consideration that most of them are already gone. The galaxy is some fifteen billion years old. If such civilizations have been emerging at a more or less constant rate for all that time, for every one around today there are a thousand that flourished in the past and since have died out.

That's a lot of information to have lost. And most of it *would* be lost, in my view, because even if the ruins of such a civilization remained on the planet, or the civilization went off the air but continued to thrive in silence, the distance between the stars is so great that we'd not usually be able to go there and find out about them.

So what do you do if you are a civilization and that knows this? Let's say that there are a hundred or so civilizations at a given time in the Milky Way, and they want to communicate with one another, and they have an interest in preserving their history. They don't want their cultural record to be hostaged to the fortunes of any one civilization. What do they do?

The answer, it seems to me, is to construct an interstellar *network*. The network handles communications traffic, and it has a memory—it keeps a record of all the data it has passed along. This is the idea that I'm going to discuss with you later in this talk.

Imagine that there are, then, many inhabited planets scattered around the galaxy that are in communication with one another. According to the Lonelyhearts scenario, each of these worlds maintains direct communication with each of the others. That means that each world must keep antennas aimed at all the other planets, sending questions and answers back and forth and so on, and no world is privy to the communication going on among the others. That approach is expensive. It's inefficient. It's as if you installed a separate telephone line going from your house to each person you ever phoned.

A network is much more efficient. If interstellar communications are handled not directly but via an automated network, each world need only keep in touch with one local terminal, through which it can access all the other data flowing among all the other participating worlds. These terminals could be established on asteroids orbiting stars at strategic points around the galaxy. The metals required to construct the terminals' antennas could be mined from oars in the asteroids, and their computer memory banks fashioned from the silicon that's also widely available on asteroids (silicon is the second most abundant material in the universe).

It's not terribly difficult to start building an interstellar network. You construct a small, smart probe—its mass probably about that of a grapefruit —and dispatch it to another star system. Once there, the probe lands on a suitable asteroid and sets up housekeeping. It builds antennas, fashions the memory it needs, calls home, and gets on with the business of interstellar communication. Most importantly, it makes a replica of itself—a duplicate of the original, grapefruit-sized probe—and sends it trundling off to another, remote star system in a suitable location.

All this may sound futuristic, and it is, but establishing a network need not be very expensive, since the society that sets the process in motion needs to invest solely in the first probe and the fuel required to send it to another star. The fuel bill represents most of the expense, but it need not be exorbitant, because the probe can go slowly: A probe that went only 100 times faster than the Voyager spacecraft could get to an average neighboring star in less than 1,000 years.

Each terminal on the slowly-expanding interstellar network that I'm describing is chartered to do three things:

First, *broadcast*. Start searching for more inhabited worlds. Broadcasting an acquisition signal is tedious, endless work, much better handled by automated stations than by biological creatures like us.

Second, *relay* all traffic to and from communicative worlds in your part of the galaxy. That's the bread-and-butter business of the network. Worlds subscribe because it makes interstellar communication easier to handle.

Third—very important—store a *copy* of everything it relays. As I will try to make clear, it is this archival function that gives the network its greatest long-term value. The advent of computers has shown us that archives need not be big, massive things like the Library of Congress. A study by the physicist Richard Feynman of the technical limits of data storage technology indicates the contents of all of the books on earth—our entire cultural memory, in other words—can in theory be stored in an object the size of a period at the end of this sentence. The tip of a pencil, if you like. If so, an asteroid the size of a city block would be suffice to contain a record of everything communicated on an interstellar network for a very long period of time.

Fourth, each terminal *distributes* data to all the other terminals. This makes the network a holographic system, one in which most data are available at every node.

Finally, the network is chartered to go forth and multiply—but to be reasonable about it. You don't want to send a cancerous probe endlessly duplicating itself. But if it does make a couple of copies and send them out to other stars strategically located around the galaxy, it can keep building itself so that it meets whatever demands of traffic it has. If it finds an emergent world on the other side of the galaxy, the network itself takes care of setting up a terminal there and searching for more planets in that area.

Such an interstellar network solves the Lonelyhearts problems that I mentioned earlier.

You'll recall that in the Lonelyhearts scenario each species has a problem with fear of transmitting, because they don't know what's out there. That leads to a possible scenario in which everybody's listening and nobody's sending. The network, however, has no such problem. It can transmit fearlessly. If it happens to reveal its presence to the most pathologically homicidal species in the galaxy, which responds by dispatching an infernal machine that destroys the terminal, that fact doubtless will be of interest to subscribers. But it will have imposed little damage on the network itself, since most of the destroyed data will already have been copied to other terminals.

The holographic nature of the network solves most of the problems arising from the long Q&A times imposed by the vast distances likely to separate most inhabited worlds. In using the network, one is accessing not remote planets but a library of cosmic history, the nearest terminal of which might be only a couple of hundred light years away. And that terminal would contain information covering a great deal of galactic history.

If such a network existed, what would the traffic on it be like? We're inclined to think of messages as simple dot-and-dash affairs, but once you start to think about hundreds of civilizations communicating by network over many years, the spectrum of potential messages starts to look a lot richer. What I think you would see are things more like computer programs—that is, sets of data that when downloaded to your computer can recreate the environment of the world that sent the program.

Interactive programs are inherently unpredictable, so the experience of immersing yourself in such a program would be a lot like real life. We in California call such programs "virtual reality," or VR—known on the East

Coast as "artificial reality"—and they offer a vastly superior alternative to getting on a space ship, saying goodbye forever to your loved ones, having yourself frozen and setting off on an expensive, centuries-long journey to another star. You don't normally need to *go* to other planets, because virtual reality offers a way to immerse yourself in alien environments without leaving home. All you require is a VR message from that world, or to have sent an automated probe there that transmits back the relevant VR data.

This technology, incidentally, has the potential of democratizing the unmanned space program. If, for instance, you send a probe to digitize Mars, you can load the resulting program into your computer and "go" anyplace that the probe went. Nor are you limited to going the same way the probe went. You can explore Mars, look at it in different levels of resolution, take your own trip. VR promises to open up whole worlds to high school classrooms and so on. It's very interesting technology.

To sum up, I expect that if we do eventually succeed in receiving an intelligent signal from space, it may come not from an inhabited world but from an automated node of a galactic network. Via this network we could learn, not only about planets inhabited today, but about many others that arose and declined earlier in the history of our galaxy. To forge such a communications link would do more than to establish that we are not alone. It would open the doors to vast libraries of knowledge, the study of which could transform our own culture.

The potential benefits of SETI have, I'm afraid, sometimes been portrayed rather unrealistically. One astronomer imagined that the aliens might provide us with a cure for cancer, another that they would teach us how to live together in peace and harmony. Well. we already know how to live together in peace and harmony, but we don't do it. I doubt very much, humans being what they are, that a message from space will serve to alter us totally—as in the Monty Python skit about the joke so funny that it kills. We're not, I think, going to receive some epigram that if shouted in the streets will change everybody and turn our world into a utopia. So, what are we doing, really? Why bother to search for companions among the stars?

I'm beginning to think that whenever we engage in communications, we are acting on what amounts to a genetic imperative. Ludwig Wittgenstein once said that "the world is the totality of facts, not of things." What Wittgenstein means, in part, is that none of us has direct experience of a thing. This doesn't mean that things don't exist; it means that the concept of "things" is a derivative concept, one based on facts. What we have are the data of our perceptions. The processing of these data goes on in the brain, from which we deduce the existence of objects in space and time. The basis of all concepts, then, is not objects and space and time, but *information*. And information can be sent out across the universe.

From the perspective of quantum physics one holds that nothing can be said to exist unless data can be obtained that imply its existence; as Neils Bohr used to say, "No phenomenon *is* a phenomenon until it is an *observed* phenomenon." The act of observation, in turn, consists of two parts. First, we "collapse the wave function": This means that data are collected, as when we intercept the light from a star. Second, there must be an irreversible act of *amplification* that *records* the observational data, as when the starlight darkens silver grains on a photographic plate that in turn is examined by an astronomer.

The idea of amplification is intriguingly open-ended: It implies that for an observation to qualify as an observation, the data must not only be recorded but also be *communicated* somehow. Communication, in other words, is fundamental to physics. It's also fundamental to all biological processes. Biological evolution, for instance, involves communicating genetic data down through generations. Indeed it is not going too far to say that communication is essential to the very concept of existence.

Now pull back your frame of reference, and consider the totality of everything we've ever done as a human species in the broad context of cosmic time. Once we perish—and we *will* earish sooner or later, you know; we'll either become extinct or we'll change into something unrecognizable—the question arises of whether, philosophically speaking, we ever existed. If we go through our careers and *never* make contact with another civilization, then by the terms of our own philosophy of science we shall not have existed. We shall have become an unobserved phenomenon, a nonentity. If on the other hand we do communicate—leaving a record of ourselves with the inhabitants of other worlds, or in the memory banks of an interstellar network—then we will continue to exist so long as scraps of that record make their way down through the long corridors of cosmic spacetime.

The attempt to communicate with extraterrestrial species has to do not just with the question of whether they exist. What we are asking, in a real sense, is whether *we* exist.

# Questions and Comments

**Fuller:** We'll start with questions and comments from the panel—Professor Morrison.

**Morrison:** As an old player in this interesting and so far baffling game, I'm delighted to listen to Tim Ferris, who has good ideas to be looked at from a still wider perspective. However, I would like to ask him a question because I'm interested in practical assistance from any new entrant in the game. In the communications world there is usually a division in the effort to communicate anew with another information source. It is not a continuing conversation, but I call up somebody. The best example is on your computer. You start punching the telephone numbers in and the modem will ring somebody up and before anything happens there's an acquisition period. It is called the period of acquisition in which an exchange is made, of course it's easy for us, but in any case it would have to be some kind of a search adjustment to see what kind of thing was coming. It's called a position of acquisition. If you don't hear the signal, there's no way to go any further with the whole thing. So, in the SETI business, since we've never heard any signal, we don't propose to send any for a time like the transit time. We feel we should occupy about one transit time in trying to acquire for the first time a signal from anything—be it network, artificial, natural, from anywhere—and we don't let anybody do it, the way we're doing it.

Do you differ with that? I think you were looking forward into a period when dozens of acquisitions have occurred and people are trying to see how best to handle the embryonic network.

**Ferris:** If I understand the question, you're asking if the strategy for detecting an acquisition signal should be altered . . .

**Morrison:** . . . Modified, if you consider the network possibility.

**Ferris:** Not very much, in my opinion. The only real difference would be that if you take the network concept seriously in designing a SETI search strategy, you don't want to rely heavily on making narrow assumptions about where Earth-like life forms are to be found. That's because what you're looking for—an acquisition signal intended for emerging worlds like ours—may be coming not from an inhabited world at all, but rather from a network terminal. But as you know, most people in SETI try not to rely too much on assumptions of that sort anyway.

It's rather like doing research in a library. When I'm working in a library stacks, about once every hour or so I get up for 5 or 10 minutes and just walk around and look randomly at the shelves. One often finds something interesting that way. Similarly, in SETI you want to randomize at least part of your search—to free it from imperatives about looking for Earth-like planets and Sun-like stars.

Another implication of the network hypothesis is that it suggests that signal acquisition might be a little easier than otherwise, since the network has the patience and the energy to make itself conspicuous.

**Morrison:** You know, the ideal case I would like you to consider is the one that we've been talking about some but it's daunting to the investigator and that is here we are looking, face on practically, 30 degrees away from face-on, the Andromeda galaxy, M31, which has got to be as close as you can come to another Milky Way. It's got 100 billion stars. It's got all the F G [i.e., sun-like] stars you want. It's got all the possibilities. If your 100 civilizations at a time are there, they're all there, or their network is sending, and surely they're not ignoring this place, which they can see also face on, and they say to themselves, "there must be life there," and so on. But there's this terrible disaster that there's a 2 million year, a 4 million year round trip time which we can't face calculating.

**Ferris:** Yes.

**Morrison:** We just don't know how to take our human activities and our human equipment and ask what could we do to imagine somebody sending to us who isn't expecting anything to come back, who cannot in the nature of thinking get anything back before 4 million years have elapsed.

**Ferris:** No, a biological being probably will not have any reason to do that, but a network itself could.

**Morrison:** Yeah it could. But it has to last a long time. It's not the question of biology, even electronics doesn't last 4 million years. There's no 4 million year old piece of electronics going.

**Ferris:** One could imagine an intelligent network engaging in inter-galactic communication, precisely because it can afford to wait that long. But the kind of signal path it would use to communicate with another network in another galaxy would not be the sort of thing that would be easy to eavesdrop on. To intercept a data path between two networks in two galaxies you'd probably have to be right in its path.

**Morrison:** No, I think it would be very easy to eavesdrop on. That's what I don't understand. Why would it not be just as interested as I am in getting an easy answer?

**Ferris:** It might be.

**Morrison:** I think it would. I know exactly what signal it would send, but what I can't figure out is how it managed to live for 4 million years to do it. It has to keep regenerating itself. It has to be biological in nature. I'm strongly of the opinion that artificial virtual intelligence is enormously exaggerated in its impact, because biology has shown the way. Biology is far more complicated and far more enduring, far more diverse, far more adaptive than any of these programs which only mimic biology at a very small rate. And I think biology is the way, man. And I don't think that anything in southern California can beat it.

**Ferris:** I think artificial intelligence is better than biology for the kind of things that I'm talking about a network doing. The network doesn't get bored; it doesn't get solipsistic, and it doesn't require the kind of support systems that biology requires. Its function is largely archival, and for that purpose it's a good idea to get it out of biological hands.

**Morrison:** I wonder if that's true. I mean, it has to have a hell of a transmitter. It was not archival at all. Just the power to send a narrow band signal from the Andromeda here to a very small region in our galaxy is a huge amount of power which is nothing like archival.

**Ferris:** It calls for an ambitious network.

**Morrison:** It calls for a very ambitious library.

**Ferris:** But what's the network going to do, otherwise, during those long dry spells when little is going on locally, except to try to get in touch with other networks? I don't pretend to know its intentions. but I do know of no barrier that prevents an artificial system from becoming intelligent. It doesn't have to be like human to be intelligent.

**Morrison:** No, no, of course not. I entirely accept that. I'm just not so sure it's advantageous. Your argument somewhat depends upon the fact that it's a lot easier to imagine this than a humanlike or a post-humanlike society continuing, and I'm highly skeptical of that. You know, it might be some relic thing that sends out that yes we were here and it sends out the first 10 chapters of its constitution, but who's going to listen to that unless there are many of them and again that would be very dull too. An archival thing that has this great cube of all information which died 4 million years ago is not going to be very interesting to the people who get it anyhow.

**Ferris:** It's not interesting if it dies. But it has to die all over the galaxy at every station. It's hard to find a way to kill it, really.

**Morrison:** OK, well said.

**Fowler:** Is there a frequency band which we would not be able to receive if it was sent by someone? Is there anything, do we cover everything with our radio and TV?

**Ferris:** Phil can answer that better than I.

**Morrison:** Well, the principle that most people worked on is that you look for the place where the signal-to-noise ratio is advantageous. There are a lot of those. You can plot signals/noise over all frequencies you can imagine. And then you see which ones will come through and which ones the medium have failed to attenuate. And that gives you a fairly narrow band that's 10DB better than anywhere else. That microwave band, and that's where we look—between 1 and 10 gigahertz roughly.

**Ferris:** Within that band there are a lot of frequencies, and you want to take into consideration shifts in radio frequency induced, for instance, by Doppler shifting of a rotating planet.

**Fowler:** But you're not answering my question. Is there a band that some other civilization might be using that we would not be able to detect and don't say that it hits the atmosphere because we can get above the atmosphere.

**Morrison:** Well, of course, neutrinos for example.

**Ferris:** You're speaking solely of the electromagnetic spectrum?

**Fowler:** No. Neutrinos of course.

**Morrison:** Sure, I mean, but we don't patrol all frequencies adequately. We can't afford it. But we patrol the best and that guess at what is the best will gradually spread as we get better and more affluent.

**Ferris:** One can take a brute force approach, using multi-million channel receivers, but you are still making guesses as to where you ought to be listening.

**McMullin:** At one time, in more optimistic days, a number of people, Carl Sagan among them, computed that the likely number of places in our own galaxy inhabited by communicating intelligences was of the order of 100 million. That was the figure that he came up with. Now I think people who work in this area are much less optimistic today, and I guess I just have a general question. If one would want to make a very rough estimate of the likely number of such centers of possible communication in our own galaxy, I take it it could be anywhere between zero and 10 billion, anywhere? There's no reason to prefer one figure over another?

**Ferris:** Well, in the absence of hard data the estimates naturally vary widely. An argument that I think is <u>not</u> well founded, one that's been used against SETI recently, is to look at all the twists and turns of human evolution and conclude that it's unlikely that the same thing has ever happened on any other planet. That's flawed reasoning. Consider this panel here today. The odds that this particular group of people will find themselves at this table at this particular time are so small that it shouldn't have happened anywhere in the Milky Way Galaxy. But here we are.

**McMullin:** On the other hand, the opposite danger is to take the theory of evolution as predictive, which in fact Sagan did and that of course is wrong too.

**Ferris:** Yes, I think that may be an error. What one used to hear in SETI circles was intelligence has great adaptive value, great survival value, so as soon as it shows up, it wins. But the counter argument is quite convincing, too: If intelligence is of such great value, how come it didn't show up sooner? How come it took 4 billion years before any "intelligent" creatures came along?

**McMullin:** And you have made a rather persuasive case to say that it might not last long.

**Ferris:** SETI acts as a mirror: Because it asks us to estimate the average lifetime of technological civilizations, it obliges us to contemplate our own fate as the one technological civilization we know of. We wonder how long we can carry on this experiment of running a technologically advanced world without screwing up the world so badly that we can't live here anymore. The issue is not the earth. Planet Earth will get along fine. The issue is our own survival <u>on</u> the Earth. If we insult the organism to the point that it rejects us, then we're the ones who get rejected. It's a perfectly honest, fair situation.

**Morrison:** I'm less inclined. There are no biologists here so I can't really engage in the argument very much on a technical basis but you say, well, it took okay it took 4 1/2 billion years, but it really only took 1/2 billion years maximum. It's tenfold better than that because surely it requires metazoans and they didn't appear for a very long time. I would say that you could well argue from the same argument that it's much more difficult to have life at all time, biochemical bacteria, than it is to have intelligent life. In fact only 10 percent of the time to go from bacteria to radio astronomies. So I would argue okay. And our argument was never that it was an inevitable thing, but that if you stretched the time scale, there's no synchronization system that we can see. So if you stretch the time scale even stars in the sun's generation have probably sent somewhere between, let me imagine, between 2 1/2 billion and 5 billion in this evolutionary phase and if they've spent 2

1/2, there's not much chance they spent 5 billion, it's a billion years ahead of us, so there's a huge spread. It was the low, it was the insensitivity of the result to the time scale that is, I think, the basis of this argument. Not so much the inevitability of the development. It is adaptive. But the thing is. The biologists always say, well it would never happen. The lung fish might have missed a vote or whatever. But wait another 200 million years and what other species will come. If you look back at the whole thing, I'm not so sure that evolution is not predictive. I think it is predictive. It's only probablistically predictive. It's not in any way a certainty, but I find it very unlikely to look at the systematic growth of the mass and number of organisms and even at the biomass given all kinds of catastrophes that have happened and still it does have a very strong monotone appearance. Life fills up land and there was nothing whatever on land until early Devonian or a little before. That's quite a big change.

**Ferris:** It's very difficult to do science when you only have one example of a given phenomenon, and we have to date but a single example of a living planet. The optimistic SETI scenarios make fascinating reading because they're so imaginative. For instance, some people in the field argue that we ought not disregard short-lived, hot stars because there one might find accelerated evolution. On a planet orbiting a giant B star, for instance, life basking in its flood of energy might evolve a hundred times faster than on the earth. It's a lovely idea. I have no idea if it's possible.

**Harrison:** It took 3 or 4 billion years to put the cell together and then less than a billion years for unicellular life to become multicellular. And now the homonids have evolved into human beings in less than a million years and once intelligence emerges, things are changing rapidly and I, therefore, wonder whether this whole discussion is on a sufficiently imaginative level. If there is life out there, it is probably on the evolutionary time scale as intelligent life is going to be millions of years different from our level and we cannot conceive what human beings will be in a thousand years. We are discussing this whole subject in the context of primitive science or primitive intelligence which is what we have and creatures that are a million years more evolved in intelligence will not be looking at the universe in the way we do, constrained by what we call the laws of nature. The situation will be totally different. In fact, life out there may have forms that we cannot recognize as living and they, if there's extraterrestrial life out there, it's what the ancients worshipped as the gods of the universe. They are so unimaginatively advanced in intelligence. They're not little green men.

**Ferris:** One question that arises is that whether such advanced species would have any interest in contacting just one more emerging world in the suburbs of the Milky Way Galaxy. If they're aren't interested, then our only chance would be to see indirect evidence of them—signs of huge engineering projects, for instance. You know, if someone's building something by melting down 100 stars near the galactic core, we ought to be able to see signs of the mess that's being made at the construction site.

**Harrison:** But that's the kind of extrapolation that we're making, a bigger core of engineers. You see. Are we still building dams on a bigger scale on this kind of thing. But highly evolved intelligence may not think or work that way.

**Ferris:** No, and the longer you look at information theory, the less big, expensive, physical modes of travel—like interstellar spaceflight via starship—make sense as a main occupation for a society.

**Fuller:** That gets us to a question from the audience which says, what would constitute an intelligent signal from outer space? Do the SETI scientists have that well defined? What would convince us that we have received such a signal?

**Ferris:** Pretty much all the radio noise that radio astronomers study comes from recognized sources, like radio galaxies, interstellar gas clouds, pulsars and so forth. Someone who is intentionally broadcasting a signal would be well advised to make sure the signal does not resemble such natural sources.

**Morrison:** To add a more technical constraint, you do two things if you send a narrow-band signal. You distinguish it from all known natural signals, which are never narrow band, and you save energy. This dual purpose does seem to make it an attractive acquisition signal. You put all your energy into one frequency basket, or a few baskets, and you use a lucky, frequency decided on, and it gives you a very attractive feeling that 300 watts will reach nearby stars.

**Fuller:** One more question from the audience. Do you regard so-called sightings of UFOs as any evidence of extraterrestrial life?

**Ferris:** No. I was interested in UFOs when I was a kid, and I looked into the field closely in those days. But the UFO stories fall apart when you get down to cases. I've seen unexplained lights in the sky myself, but you are not justified, every time you see an unexplained light in the sky, in concluding that it must be an alien starships. That's an answer, all right, but it's an unsupported answer. It's not science but the pseudosciences that have an answer to every question.

# EARLY NUCLEOSYNTHESIS IN AN INHOMOGGENEOUS UNIVERSE

## WILLIAM A. FOWLER

My Nobel lecture in 1983 addressed the question of the quest for the origin of the chemical elements in our observable universe; quite simply it was argued that isotopes of hydrogen and helium were produced from protons and neutrons in a homogeneous and isotopic early universe commonly called "The Big Bang." Still heavier elements and their isotopes were synthesized from the primordial hydrogen and helium by nuclear processes in stars, nova and supernova. It is still believed that the great majority of the elements beyond helium are produced in these astrophysical circumstances. However, since the suggestion by Alan Guth in 1981 that our early universe experienced an enormous growth in size or *inflation*, it has come to be believed that a small fraction of the heavy elements were produced shortly after this inflation. This explained among other things why the earliest oldest stars observed by astronomers show evidence for small but significant amounts of heavy elements in their spectra which could not have been made in stars because there were not any previous ones to these oldest ones.

Studies of this new development was greatly enhanced when Edward Whitten in 1984 showed that the universe after inflation would be inhomogeneous rather than homogeneous. Our observable universe, expanding as Edwin Hubble found from his red shift measurements in the 1920s, could be thought of as an expanding bubble of matter as we know it, into an otherwise steady state universe consisting of extremely high density stuff called vacuum matter, for want of a better term. This vacuum matter corresponded to Einstein's cosmological constant and to Friedmann's cosmological term in his equations governing the behavior of our observable universe with a finite but rapidly increasing radius. Remarkably enough this vacuum matter exerted negative pressure and thus can be thought of as the cause of our expanding bubble.

The matter in our expanding bubble was inhomogeneous and the early quark-gluon plasma transformed into a hadron gas consisting ultimately of high density proton rich regions immersed in a neutron rich sea. Nucleosynthesis in the neutron rich sea permitted highly charged, heavy nuclei to be synthesized since neutrons are neutral with zero electric charge. Coulomb repulsion prevents charged protons and alpha particles from amalgamating with highly charged nuclei. There is no Coulomb repulsion for neutrons.

Many authors, including myself and my collaborators, have contributed to what is by now a fairly clear picture of nucleosynthesis in an inhomogeneous universe. Experimentalists have studied in the laboratory some of the many additional reactions which took place in the neutron rich sea which was also rich in deuterium, tritium and helium nuclei. This past decade has been a very exciting one in contributing to our knowledge of the early history of the universe you and I inhabit. There's still more to be done and that is mainly what I will talk about as I go to my slides.

Before I go to my slides which are going to be rather dull, I'm afraid, let me tell you a story which I think is appropriate to tell at Gustavus Adolphus College. It's about a priest and a member of his congregation who I will call a practitioner. The priest and the practitioner went to play a round of golf one day. They got to the first hole and the priest teed up his ball and took a great swing at it and missed it completely. He exclaimed, "God damn it; I missed." The practitioner rushed up to him and said, "Father, should you be using such words even on the golf course?" The priest just shook his head and they went on to the second hole. The priest teed up the golf ball, swung again, and the same thing happened. Once again he said, "God damn it; I missed." This time the heavens opened and a great roll of thunder and a lightening bolt came down and struck the practitioner dead and a great voice from heaven said, "God damn it; I missed." Well, I guess we'll go to the slides now.

It is the inflationary universe which I will be talking about. For those of you who are interested in further reading, Guth's first article had some errors in it but they were all cleared up by Guth and Paul Steinhart in an article in the Scientific American in 1984. The whole point is that the early universe before inflation was very small with all parts in causal contact and thus at the same temperature. That led to the fact that the cosmic background radiation is at the same temperature everywhere today, as observed, through better than one part in ten to the fourth. Previously in our early ideas of the big bang, we really had no way of understanding the constancy of the cosmic background radiation but when you start out with something small enough, it was all in causal contact, thermal equilibrium was established and that continued during the expansion. The early exponential expansion during 10 to the minus 32 seconds, led to a flat Euclidean universe with a curvature parameter, which appears in Friedmann's equation, thus going to zero. If one expands something like a sphere and looks at the curves of longitude on it, at first they all look like they are curving also, but if you expand this sphere and look at just a very small part, the lines of longitude get closer and closer together and the curvature which is described by the constant k in the Friedmann equations goes to zero. That's one of the basic things which comes out of inflation. I've always liked this because it makes Friedmann's equations much easier to solve than when k is positive or negative. Thus there will be a universal expansion rate in describing the expansion and I will take A as a distance scale factor.

Hubble's constant is equal to the time derivative of A divided by A. Friedmann's equation yields values for $8\pi Gpt/3$ where G is the gravitational constant, and pt is the mean matter density in the universe. This density, pt, includes real matter and the so-called vacuum matter, which is assumed to fill all space uniformly. If one sets the curvature parameter equal to zero, as discussed previously in connection with inflation, then one obtains

$$H^2 = (\mathring{A}/A)^2 = (8\pi G/3)pt = (8\pi G/3)\ (p + pv)$$

This series of equalities can be simplified by setting

$$\Omega = 8\pi Gp/3H^2 \text{ and } \lambda = 8\pi Gpv/3H^2$$

so that

$$\Omega + \lambda = 1,$$

which for

$$\lambda = 0$$

yields

$$\Omega = 1.$$

This $\Omega$ can be expressed as

$$\Omega = \Omega_b + \Omega_e$$

where the subscript b designates baryons or ordinary matter while the subscript e designates exotic particles such as neutrinos, photinos, axions,

Higgs bosons, etc., etc. The standard Big Bang production of $^2$H, $^3$H, $^4$He, and $^7$Li abundances has required $\Omega_b \approx 0.1$ as noted by Robert V. Wagoner, William A. Fowler and Fred Hoyle as early as 1967 in Ap.J. 148, pp. 3 to 49. With $\Omega = 1$ these results yield $\Omega_e \approx 0.9$. This large amount of exotic matter has never appealed to me. I have enough exotic things in my life without having so many particles around that are exotic in great numbers. Could $\Omega_b = 1$? My answer and that of other people is yes from studies made in the 1980s. For example refer to R.A. Malaney and W.A. Fowler, American Scientist 76,472 (1988).

Now, more about the inflationary universe. A wonderful thing, both in classical mechanics and in general relativity, the total energy of the universe, or of anything for that matter, is equal to the kinetic energy plus the potential energy. The kinetic energy is just one-half mass times the velocity squared and the potential energy is the gravitational constant times any mass squared over the separation distance. The curvature term in Friedmann's equation creates difficulties in this regard. However with the curvature equal to zero, one finds that the total energy in the universe is zero and it has always pleased me that in this general picture, the conservation of energy is observed. So inflation reduces the curvature parameter to zero and thus the total energy of the observable universe is zero. Energy is conserved in the universe that we inhabit. There is no problem in the conservation of energy in the expansion after inflation. The expansion of the universe, follows the conservation of energy, in other words, or is driven by the conservation of energy.

Consider other consequences of the inflationary model. First of all, I've tried to convince you that the total energy of the universe is equal to zero. The rest mass energy plus the kinetic energy of expansion minus the gravitational potential energy is zero. And then you can show if you again go back to Friedmann's equations that time in the universe is governed by a relation between the age of the universe and the Hubble time. Hubble's, law yields velocity proportional to distance. If you bring the distance over on the side of the equation where the velocity is, velocity divided by the distance is time and that we call the Hubble time. It can be shown that the age of the universe is two-thirds of the Hubble time when the curvature term in Friedmann's equation is zero. I must note that there is a great deal of controversy about this simple "time" relationship. I think it is true and many of my colleagues who had worked on the problem think it is true but I want to emphasize that there are skeptics who would challenge even this simple relation which holds for a zero curvature universe.

Let's look at a possible scenario based on the inflationary model which is very oversimplified. We think we're at the center of things and we can look back in time as we look out in distance and we can look back to the period when the universe was opaque. It was at the end of that period that the source of the black body radiation occurred. It started out way back then at about 4000° Kelvin and now due to expansion it has come down to 2.7° Kelvin as is accurately measured. When you expand any gas, it cools. Now the observable horizon is the velocity of light times the Hubble time, but the true horizon can be much larger. In fact, if you want to dream a bit, you can think that there may exist other universes immersed in the "false" vacuum. The false vacuum is equivalent to a large cosmological constant. It obeys Friedmann's equation. No question about that. It just has a large cosmological constant and what we hope is that here right around us that the cosmological constant is zero. So that's a possible scenario. By the way, these other universes have to be at great distances from us and each other because in the roughly 10 billion years of the age of our universe, we don't want these other universes overlapping. The consequences would be spectacular and disastrous not only for the overlapping universes but for all other universes including ours.

In the very beginning of our observable universe, the temperatures were so high that the protons and neutrons were broken down into their constituent particles which are called "quarks." As things cooled a quark-hadron phase transition occurred. Three quarks fused together to make hadrons. There were some heavier hadrons than the baryons which gradually admitted gamma rays or various other things and came down to those which make up the stuff like us. We consist of the lowest mass-energy type of the hadrons. Many authors have shown that the early quark-hadron phase transition makes nucleosynthesis in a closed baryonic universe in fair agreement with observations on the primordial abundances of deuterium 2, helium 3, helium 4, lithium 7, beryllium 9, and a small amount of the elements with atomic number greater than 12. The key is that the baryon density fluctuation leads to a neutron rich region in which part of the Big Bang synthesis takes place. There is no Coulomb repulsion between neutrons so in the neutron rich region nuclear reactions can involve more heavily charged nuclei and produce nuclei with atomic numbers greater than 12. In fact, the reactions can proceed all the way up to uranium and thorium even though only a very small number of heavy nuclei is made at this stage. Those who follow my line of thought think that this is where the small amount of heavy elements that astronomers see in the very oldest stars, stars in the oldest galaxies, came from.

I now turn to what Robert A. Malaney, now at Livermore, did with me looking over his shoulder more or less. We decided to investigate Big Bang nucleosynthesis not only with Omega equal to one but with the Omega being made up of bariums that is, $\Omega_{b=1}$. Following the quark-hadron phase transition, there existed two distinct types of regions, again an oversimplification. There were high density regions which were very proton rich. They were much like the whole universe in the standard homogeneous model. In these regions there are about 6 times as many protons as neutrons, because neutrons are heavier than protons. In the ratio of protons to neutrons, there is an exponential term that contains the mass difference to a negative power so protons turn out to be 6 to 7 times as probable as neutrons. That is the case in the bubbles where all the matter started. But neutrons diffuse much faster than protons. When a proton is moving it can collide with another charged particle and be scattered usually with some energy loss. Neutrons are not scattered by charges so they "waffle" around on their own. Finally they escape from the high density bubbles and produce a low-density region which is neutron rich and which surrounds the proton rich bubbles. There is a different nucleosynthesis in the two regions and you not only get the primordial abundances of deuterium 2, helium 3, helium 4, also lithium 7 in the observed primordial amount if back diffusion of neutrons destroys beryllium 7 in the proton rich regions. I'll say a little bit more about that. But the most important thing is that a small amount of heavy elements, mostly from A equals 12 up to about 60, is made in what is called the r-process which also occurs in stellar nucleosynthesis. The "r" comes from rapid in the rapid neutron capture process. We now see that r-process nuclei can be produced in first generation stars. The neutrons will produce nuclear reactions all the way to thorium and uranium and then fission occurs. Out of that fission there emerge two nuclei where before there was only one. So, the whole thing works beautifully in that you get more light nuclei to which you can add neutrons and so forth and so on. These nuclei form the seeds for the s-process in first generation stars. The lower case "s" means the neutrons are captured slowly, much more slowly in stars in the s-process than in the early universe.

We can go on with this. In this over simplified scenario there is a proton rich bubble in a neutron sea. It turns out that the neutrons must not be used up too soon. That is all well and good because it turns out the bubbles originally are quite dense and have very little surface so there is very little neutron penetration. Then as the expansion goes on the density in the bubbles gets lower so neutrons are not scattered so much. In the expansion the surface area gets greater so the neutrons in the neutron rich sea can reenter the bubbles. That is all to the good because in these proton rich bubbles one of the nuclei produces beryllium 7. Beryllium 7 has 4 protons and 3 neutrons. That is the nucleus which occurs at mass 7 in the proton rich region. Four protons compared to 3 neutrons. The stable form of mass 7 is lithium 7 which has 3 protons and 4 neutrons. When Malaney and I first made our calculations we had a "terrible" result because too much beryllium 7 had been produced and finally the beryllium 7 decayed by electron capture to lithium 7 so too much lithium 7 was produced. The solution at this stage of the game when neutrons came back into the bubbles, hit the beryllium 7, changed it into lithium 7 plus protons but with the protons in the bubbles the lithium 7 hit them and went to two helium nuclei so the result was not too much beryllium 7 and ultimately not too much lithium 7.

One of the things that pleased me in this work was that the inhomogeneous neutron rich regions required the study of many additional nuclear reactions in the laboratory which we didn't require in the old homogeneous universe. So in the inhomogeneous universe, the neutron rich region, brought the necessity for studying, in the laboratory, many new reactions not previously required. There are quite a few of these new reactions and that has been one of the things that pleased me most that some work that we were doing theoretically could lead to work to be done in the laboratory because by this time work on stellar nucleosynthesis in the laboratory had pretty well petered out and so this was a new shot of life for Ralph Kavanagh and Charles Barnes and others in our laboratory and other places.

In the old homogeneous universe one obtained , deuterium 2, helium 3, helium 4 and lithium 7 with the observed abundances when the density was about $3 \times 10^{-31}$ which is much less than the critical density at $6 \times 10^{-30}$ g cm$^{-3}$. So, in the old point of view, in a homogeneous universe as many other people have found, the density of the universe as measured from what went on in the early Big Bang or just afterward, was roughly one-tenth of the critical density. Since we had come to believe that there was something that was making up the critical density, we had to do it with all kinds of exotic particles like neutrinos, photinos, axions, Higgs bosons . . . I never liked making the density of the universe so that partly motivated me to go to the study of the inhomogeneous universe. In this case one has a lot of flexibility because there exists a proton rich region in which nucleosynthesis can take place and also there exists a neutron rich region in which nucleosynthesis can take place.

To make a long story short, one of the nice things about the inhomogeneous model is there are some quantities which can be varied. One is the fraction of the volume in the neutron rich region relative to the fraction of the volume in the proton rich region. Malaney used a computer to search for the best fits to the abundances and found another adjustable variable in the rate of the neutron diffusion back into the proton rich region. It is very difficult to calculate neutron diffusion. Anyone who's in the field knows that so all we did was say that the ratio of the neutrons in the proton rich region relative to the neutrons in the neutron rich region was the quantity $A_0$. A large $A_0$ indicates very rapid back diffusion. $A_0$ equals zero signifies no back diffusion. The calculations showed that $A_0$ greater than about 0.3 gives quite good agreement with the observations. The observed deuterium is somewhat greater than $5 \times 10^{-6}$. The calculations yield approximately $10^{-5}$ for $A_0 = 0.3$. The observed helium 3 is less than $3 \times 10^{-4}$. The calculations yield $3 \times 10^{-5}$ which does not contradict the observations. The helium 4 abundance is a problem. The calculations yield too much helium $4 = 0.25$. It depends on how serious you take the people who think they can specify the amount of helium when our galaxy formed. Originally there was some latitude in the observations. Now the abundance has been tied down to .23 and, if that's the case, the calculations are high. But this is a very oversimplified calculation and I am still working on this problem with others in our laboratory. We're trying to see if there isn't some way to get this calculated number down to a more reasonable value. In the Large Magellanic Cloud the abundance of lithium 7 runs between $2 \times 10^{-10}$ and $8 \times 10^{-10}$. The calculations for $A_0 = 0.13$ fit this range. So, in general, this picture of nucleosynthesis in the inhomogeneous universe agrees fairly well with the observations except for the fact that too much helium 4 is made and that's fun because it leaves some more work to be done.

We turn to Big Bang nucleosynthesis of the heavy elements. It all takes place in the neutron rich sea. You go very easily up to $18O$ with neutrons, deuterons, tritium and alpha particles. In addition, rapid neutron capture occurs all the way up to fissionable nuclei. New seed nuclei are produced. Moreover, fission produces additional neutrons about three per fission process. Thus fission cycling can occur. Malaney and I found that 25 cycles would do the trick and that's another arbitrary number but a reasonable one. Fission cycling produces the low abundances of the heavy elements ($10^{-4}$

solar) found in the oldest stars in the galaxy so our galaxy formed with a small amount of heavy elements produced in inhomogeneous Big Bang nucleosynthesis.

Numerous problems remain. The luminous matter consists of baryons. This luminous matter is about one tenth of the critical density and so if $\Omega_0 = 1$, then what is the dark or missing matter? I believe it is baryons but they have to be dark and that's one of the problems. There's another problem; the conventional globular cluster ages can be 16 billion years, but that is much greater than $T_0 \approx 10^{10}$ years. One way to reduce globular cluster ages is to have main sequence mass loss when stars first form and that has been studied by Lee Ann Wilson at Iowa state. In addition, Winget et al. find that the age of the oldest white dwarfs is less than $T_0$ so the conventional globular cluster ages may be high.

The epitome is we may well live in the simplest of all Einstein's universes. His curvature parameter is zero. His cosmological constant is zero. His space time is Euclidean. His universe has zero total energy and his matter is stuff like us. I think Einstein would like that; I know I do and I hope you do. I remain at work in this field driven by this hope.

# Questions and Comments

**Fuller:** We will start with comments and questions from the panel. Professor Harrison.

**Harrison:** It's always nice to ask where things originated and here is a quarter and to ask where did the metal come from. Now, the answer has been that this metal was made in a star that died before the birth of the sun. That star exploded, sending the heavy elements into the interstellar medium that then made up our own star and our own planets. So, that's one of the exciting facts of the new picture of the universe of this century that we and things around us, the elements were made out of the death of old stars before the birth of our sun. So, in the very early stage of the universe, which Professor Fowler has been discussing, the idea was that the elements were the simplest. Hydrogen, I think, became mainly helium and so the first stars consisted mainly of hydrogen and helium. The heavy elements according to the argument by the Burbidge's, Fowler, and Hoyle were then manufactured in stars. But now today we heard of this new approach and I want to ask Professor Fowler how much of the metal must I now attribute to the Big Bang rather than to the stars?

**Fowler:**  One part in 10,000.

**Harrison:**  One ten thousandth, that will be about a milligram say, or, no, a few micrograms.  And would it cover the elements that are common around us, the metals, the iron peak?

**Fowler:**  In general "yes" but please recall that only the neutron rich isotopes of the heavy elements are produced in the inhomogeneous Big Bang nucleosynthesis.  That's a mass.  It's all copper and nickel.  There's not an iota of silver.  Just like astrophysics.

**Ferris:**  You know one of the beautiful things about particle accelerators is that you can recreate conditions that existed at some point early in the history of the universe and the bigger the accelerator, the earlier and hotter and higher energy epoch you can replicate to some degree in the accelerator.  We're in a period now where there's a lot of controversy about big science and the most conspicuous target of this controversy I should think is the super conducting super collider project in Texas.  Now an argument that can be made against building such an accelerator it seems to me is to say the whole universe ran this experiment and everything in this room and everywhere else in this universe is the result of that experiment.  When we hear from a nuclear physicist who's as smart as Willy Fowler, isn't it possible to argue that well you're capable of reconstructing what may well have happened at different epochs beyond the reach of a very expensive grand accelerator.  So, do we need the accelerator?

**Fowler:**  Not for what I've been talking about.  In fact in this business, what is needed is lower and lower energies because the temperature corresponds to around 100 kilovolts or so.  In so far as all of this is concerned, we don't need a superconducting collider but then I am opposed to superconducting colliders and one of the reasons is is that they're not going to do me any good in my work in Big Bang nucleosynthesis.

**Morrison:**  Without entering into a brief for expensive machines in Texas, I would say that you depend heavily upon the implications of these early phenomena.  After all, the Whitten inhomogeneity and the Quark hadron transition and so on is very nice and it's all done by the theorists but it would be awfully nice to have some kind of touch of realism added by some experimental information.  I imagine they will try very hard in the next years to do something about Quark matter.  Before it is meta-hadronic in the infantly transient situation that they can find with especially large colliders.  So it's not divorced from that.  It's still the cosmological era, but what Professor Fowler has done is show us that even in the soar of nuclear physics here, which is not cosmological in some grand sense. If we add success from our experiment directly, there is a possibility for a novelty in our discussion.

Geller: Well, Willy, I have to say that I'm one of those stupid people that you refer to in your lecture because I don't believe that omega is one. And the reason I don't believe omega is one is that when we measure, when we look at the dynamics in systems in galaxies and we measure the masses of systems even on rather large scale, we never seem to come up with a number that large and it's especially a problem if the matter is baryonic. Now, it's true that we can't limit the matter which is uniformly distributed. But on the other hand, if the matter were baryonic, you might expect it to cluster the way galaxies and other things cluster in the universe. So, my question to you is, if this stuff is out there, where is it and how do I go find it?

Fowler: Well, it cannot be in our galaxy. The dynamics of our galaxy as Oort showed many years ago requires only about twice as much as you astronomers actually observe. But that doesn't mean that it isn't out there. I guess I didn't have time to emphasize that enough. People who have my point of view, we're in a little bit of a tough spot because we can't put the additional matter into large objects because then they'll shine and you astronomers would see them. And if you make them too small like dust, then they'll just obscure everything, so we're forced to say that this exotic matter, but which is baryons, is probably on the average in something about the size of the planet Jupiter. Because I claim in the intergalactic medium you would not see Jupiter.

Geller: But why wouldn't these Jupiters cluster together with clusters of galaxies and around individual galaxies. I mean, why wouldn't we detect them when we study the dynamics of large systems. You would expect that they would not remain uniformly distributed if they were.

Fowler: I confess that's one of the problems, but if you believe that nice agreement that we get and do explain in these observations that are made on the abundances in the very oldest stars, then you've got to hide it, but then that's got to be done in any point of view.

Geller: Yes, I agree.

Fowler: Unless, as you say, you don't believe omega is equal to one.

Geller: That's right. I don't believe.

Fowler: But then you can't believe in inflation.

Geller: Well, that's also true. I mean inflation is a beautiful picture, but it makes a prediction that $\Omega = 1$ and it's true that we can't rule that out. On the other hand, we don't find it and it's getting harder and harder to understand why we don't find it. Of course, omega, you could have a nonzero cosmological constant which is also not so pleasant, but inflation does explain some things but it doesn't tell the whole story from the point of view of someone who's actually exploring the universe. I mean, one of the funny things in this field, of course, is that we have very few real observations. I mean, we hang our hats on a very small number of quite big observations, one of them is the abundance of the elements. Another is that the sky is dark at night. Another is the structure that we observe. But we don't have a huge number of things and so we're not as well constrained as we'd like to be but I think we're becoming better constrained by knowing something about what at least the nearby universe looks like.

I guess the other thing that people have suggested for explaining these heavy elements is that there was an early generation of mass of stars which preceded galaxy formation and that these stars had rather short life times and they produced, and in the supernovi which were the end result, they distributed heavy elements which are what we observed. Now of course, you would say there are lots of problems with that model which make it unpalatable.

**Fowler:** Yes, yes. But the other thing is that we can point out something for the space program to do with all that money it's spending. When they get up there, outside of the galaxy, they should scoop up some of this stuff. I'm serious. They're going to spend lots of money anyhow; they might as well do something useful with it.

**Morrison:** I'd like to ask a question which is not as usual, a rhetorical question, it's a question, I really don't know the answer to. What is the learning on this and the two of you can probably enlighten me. The heavy element contamination in what are judged to be old stars has something of a problem it seems to me in this sense. These stars have been one inhomogeneous device, but maybe a second inhomogeneous device will make some trouble for it. It's all very well to separate the neutrons from the protons. I'm suggesting that the contamination of these poor old stars living in this dirty galaxy for all those 5 or 8 billion years got plated with some heavy material from the local situation and weren't born with that at all. The amount is so small that I don't think it affects the stellar interior and therefore it doesn't affect the overall physiology of the stars and it's quite hard to tell. It comes entirely from spectral analysis of the top gram or something like that of the star that is 100 million grams thick per square centimeter. So I wonder if that's been looked into to see if the mixing is appropriate.

**Fowler:** The stars have a convective layer.

**Geller:** They do, but I don't know how deep.

**Morrison:** But it's not very deep.

**Fowler:** I've looked into that. It's so deep that you've got to contaminate 2 or 3 percent of the star's mass.

**Morrison:** Well, that may be. That's still a big number.

**Fowler:** These convective layers are very interesting and important.

**Morrison:** Have you allowed for that possibility that it's just a contamination?

**Geller:** I don't know the answer to that.

**Fowler:** It's my understanding that the convective layers involved is a substantial amount, in the sun it's a quite large fraction. Isn't it?

**Geller:** I don't know. It's an interesting question, but I don't know the answer either.

**Fowler:** But it's a pretty large fraction. Once the temperature gets low enough that you can't transfer energy by conduction, then you've got to use convection of energy and the sun and other stars are giving off this energy. It goes out a certain distance by conduction and then the conduction ends and then you've got to get rid of it an the only way you can do it is to have the outer surface convect. Bring the energy from down deep up to the stellar surface where it can shine into space.

**Geller:** I might just mention that one of the appealing things about this suggestion of population 3 is that you can sort of compute if you look at the voids and the galaxy distribution, you can compute the energy input required to make them and it's not so different from what you need to actually produce these heavy elements. So some people have tried to make that connection between this population 3 as being sources of explosions which might drive the large scale structure that we observed, but there are of course many problems with that as there are with all the explanations that try to explain how the structure we observe in the universe form but there are these ideas that one of these problems might have something to do with the other.

**Fowler:** Oh, I agree to that, but I've never liked the concept of population 3 because it's all gone and there's no way we can check on what people are saying about population 3. The thing that I'm proposing can be found and it's an alternative, if you want to put it simply. It's an alternative to the population 3 point of view. If it's found that in the intergalactic medium there ain't any Jupiters or other things that can make up the so-called dark matter, then I give up and I'll go back and work on something else at which I'll make great mistakes from time to time. I would like to interject for the sake of many of our listeners who are not working in the field that population 3 is the term used for a generation of stars earlier than the observed stars.

**Geller:** Yes, I explained that earlier.

**McMullin:** This is a more general question. The inflation model itself is just about 10 years old and I think to the average person in the audience here, the notion of an expanding universe is already becoming somewhat familiar, but the notion of inflation is still, I think, hard to take. One is talking of an expansion here in the very much less than the first second of the Big Bang, in which you would go let's say from the size of a baseball to the size of the observable universe. That's one analogy I've seen used, which is, in less than a second, which is close to instantaneous. Now, the question I have, a general one, is this, does your work bring additional credibility to that idea or is it simply a way of filling it in, drawing consequences. If we accept the inflation model, then the following, this would follow. For example, specifically, if you accept the inflation model, how plausible does it make these inhomogeneities on which your entire analysis depends?

**Fowler:** First of all, you do have to have inflation and then you have to have the additional thing that Whitten at Princeton was the first one to point out that after inflation, you just don't have a uniform universe any longer. When it all settles down, it settles down into bubbles of matter, in an otherwise vacuum and then out of these bubbles, the neutrons diffuse out so pretty soon you have proton rich bubbles in a neutron rich sea. That's an oversimplified picture, of course. In the calculations that Malaney and I and others have done, we have to take the bubbles all the same size and we describe that complicated situation it must have been by two free parameters $f_V$ and $A_0$. But I think that something like that really had to happen. The fact that the curvature constant is zero from other considerations is what the inflationary model really gives you. If you're convinced the curvature constant is zero, then the simplest explanation, as I said, is an early great expansion which straightens out all the coordinate lines.

**McMullin:** Let me go back to the original question, perhaps. Supposing you were to meet someone who was a little skeptical about the idea of inflation generally, would you argue that your work in nucleosynthesis brings additional support to that or would it simply be a matter logically of saying if we accept inflation, then these are the results for nucleosynthesis that would follow. But would you see some indirect kind of support brought to the inflation hypothesis. Just like, for example, earlier the figures on helium abundance brought support to the Big Bang.

**Fowler:** Well, I don't know for certain but I think it does support the general idea of inflation because the inflationary epoch is essential but in a complicated way. Namely, after inflation according to Whitten, the result was not a uniform universe, but a large number of proton rich bubbles. I gave an analogy. There was a phase transition and the phase transition that we're all familiar with is the boiling of water. When you boil water, it doesn't all boil at once. Bubbles of water vapor are produced in the liquid water. The bubbles form around small impurities in the water. I don't know what the impurities were in the early universe but a number can be suggested. All Whitten suggested was that there was something like the impurities in boiling water that made the bubbles form instead of all the water boiling at once. It's a very reasonable picture and I don't whether what we have done proves it but it's in disagreement with the old point of view in which you can't make any heavy elements in the Big Bang. So then you have to make them in population 3 stars and as I've said ad nauseam, I don't like population 3 stars because we don't see any remnants.

**Geller:** Also, nobody knows how to make them, which is another minor problem!

**Fowler:** But the standard population 1 and population 2 stars can be observed and studied. Maybe what I am suggesting is even more nonsense than population 3, but I hope not in the long run.

**Geller:** Well, I don't like omega = plus one because I think it isn't fair.

**Fowler:** You're a difficult gal!

**Harrison:** I think it's a misconception that inflation requires omega equal to one and I see no conflict between the two of you. If observation shows that omega is equal to .1, you are caught into a phase transition back when curvature constant in any case was effectively zero. And so there is no conflict between the two of you provided you look into the theory and realize that inflation does not unambiguously demand that omega is now equal to one.

**Fowler:** You're right.

**Harrison:** That is a simplification for the sake of popularity of the subject.

**Fowler:** Yes, but it does require that the curvature of the observable universe be zero.

**Harrison:** We know that. Whatever you are going to say k is now the curvature constant effectively during the early universe so you can put it equal to zero.

**Fowler:** Well, inflation certainly does.

**Harrison:** Even without inflation, it's effectively zero.

**Fowler:** No, I don't believe that at all.

**Fuller:** We have a question from the audience. Would you explain your relation between the false vacuum and the initial Bang?

**Fowler:** Well, the false vacuum is all the surrounding stuff. It has the remarkable property that it generates a negative pressure. To get the total energy, you have $pc^2$ plus the negative pressure. The result is zero. I like that, because it means that in the early universe energy was conserved and for people who have worked in nuclear laboratories that's the kind of a result that you just have to have in order to be happy with what's going on.

**Fuller:** Another question which relates to one of the comments from the panel. How does your model develop proton bubbles in a universe that should be extremely homogeneous following the inflation?

**Fowler:** That's just an oversimplification. You're already making things complicated by having bubbles in a vacuum. Sure you can make the scenario as complicated as you want, but then you can't make any calculations without hiding things in a computer in such a way that you don't really know what's going on. So, we have adopted a very oversimplified picture of what the universe was like after the quark-hadron phase transition. If you make it any more complicated, then you can't make simple calculations. And whether some of you like it or not, that's a thing we do all the time in physics. We simplify complicated problems in order to be able to solve them. Isn't that right, Phil?

**Morrison:** And often we don't solve them!

**Fowler:** Yes, even after we simplify them.

**Fuller:** Another question from the audience for the entire panel. All of the Big Bang and steady state as well as Margaret Geller's galaxy mapping depends on the Hubble law using the red shift as a measure of cosmic distance. Arp and his group have presented evidence that this interpretation is not correct. Would you care to comment on that?

**Geller:** Arp's arguments are incorrect. I think there is abundant evidence. There was a long period when there was really a controversy about whether the red shift was really related to the distance as suggested by the Hubble Law and I think recently there have been many observations on large scale which indicate that is correct. One of, perhaps one of the most striking is the observation of gravitational lensing where you see a distant source, gravity bends light, so it acts like a funny kind of lense and we see distance quasars which are lensed by intervening galaxies and models can be made for these which assume that the red shift is an accurate measure of the distance and these scales are very large and these models are tested very accurately because of the data. Often you have radio maps of these which are extremely accurate and it's remarkable in fact how accurately the models match the data. So, I think this is one of the cleanest pieces of evidence that is recent. Another is that one of the reasons that people doubted that the red shift was a measure of distance is that quasars which are very distant objects which appear stellar have, they have a prodigious energy input if you use the red shift to compute the distance. However, people couldn't see. Now people believe that quasars live in galaxies and in fact people have seen the galaxies associated at the same red shift with quasars. Also, there are galaxies along the line of sight to quasars which absorb light and you can find the galaxy at the red shift which is associated with absorption line. More direct confrontation with Arp is that Arp makes certain statistical arguments and these have been looked at in detail by people who have made much better maps of the universe and of the distribution of quasars and galaxies in the universe than were observed before. And it turns out that Arp's statistical arguments are fallacious. So I think that although there was a controversy, it's I think now really not a real issue.

**Fowler:** A little personal story. Many, many years ago I had a class in nuclear astrophysics and of the students was Halton Arp and the other was Allan Sandage. And of course Sandage has done some remarkable work in extending Hubble's early work and by the way, he's just been awarded the Crawfoord Prize which is equivalent of the Nobel Prize for astronomy and astrophysics. In exams and everything, Sandage just sailed through the course and I of course gave him an A. Well Arp did not solve one problem! But I liked him. And so I thought well I won't give him an F because then he'd have to take the course over and I couldn't stand that. Then I thought well maybe I'll give him a D, but if I gave him a D, then he didn't get any credit at all. So, I finally gave him, what would you guess—a C- minus! I've always been glad that I did so.

**Ferris:** Einstein used to say in words that are engraved here and there, on a fireplace in Princeton for one, that the Lord is subtle but not malicious.

Another way to say this is that it is possible to concoct a more complicated but alternate explanation for any phenomenon observed in nature. You can theorize that. Well, I'll give you a better example. When geological strata were first discovered, religious fundamentalists argued against them on the grounds that God had deposited this evidence to make it look as if the world were older than the Bible said. To test the faith of those who might be tempted to disbelieve the Bible. Well that's a malicious act on the part of God. What Einstein was saying was God doesn't work that way. If we assume that the red shifts of galaxies are due to actual velocity, then everything fits in a very simple way and we see red shifts and light that are indubitably due to velocity in the nearby universe. One can construct an alternate theory that says there's another mechanism that makes the galaxies act just as if the universe were expanding. And it's possible to do that but it requires a malicious sort of construction to the universe to do it. So, for Chip Arp to be correct, God must be more malicious than I think the evidence shows.

**Harrison:** In 1929 when Hubble presented the distance measurements and the red shift measurements showing that they had a relationship that later became known as Hubble's law and forms the foundation about our belief that the universe is expanding. In that same year in the same volume and in the same journal Zwicky wrote a different paper proposing what is called the tired light theory to explain the red shifts and arguing that perhaps the universe doesn't expand at all, but as light travels these vast distances in the universe it grows tired. This is just a sort of expression. He offered no physical explanation but in the intervening years there have been many arguments and physical theories to try and make sense of this idea that light grows tired as it travels over large periods of time in the static universe. There are problems. Commonly the idea is that there is some medium throughout the universe that robs the radiation of its energy as it travels but this medium's got to be very peculiar. It's got to take the energy away without scattering the light. There is no known interaction between radiation and matter that can achieve exactly this effect so it would be, whatever it is, it seems that it is highly contrived and it would be a jest not by God but by nature if indeed that tired light theory turned out to be correct.

**Geller:** It's also demonstrably inconsistent.

**Harrison:** Yeh, you're right.

**Geller:** It makes predictions about how surface brightness of galaxies should depend on red shift and how the angular size should depend on red shift and those are demonstrably inconsistent with that.

**Harrison:** The current belief fits together beautifully in a self-consistent fashion with many other observations. I think it's quite secure. Although still we see proposals about the tired light theory, it doesn't die.

# WHERE THE GALAXIES ARE

## MARGARET GELLER

One of the things that has always been remarkable to me and wonderful about working in cosmology is that nearly every person has some opinion about what the universe is like, even if they've never studied it. I think if you went to Harvard Square and you asked people what they thought about the universe, many people would have an opinion. I teach a fair size undergraduate course at Harvard for nonscientists. On the first day I always ask them to vote on whether the universe is finite or not, and the winning vote varies from year to year. This year, by 2 to 1, the class preferred the universe finite. We'll see if they still think that by the end of the semester. But I ask the students to explain to me why they preferred one view over another. And one student who preferred the universe finite said, "Well, then I'm more significant relative to the universe."

One of the most daunting things about working in cosmology is the sheer enormity of the universe. And today, my task is to tell you about what we know about the universe and about this small fraction of it that we've mapped. This project is a story of a fifteen-year collaboration between me and my colleague and friend, John Huchra. So we, and all our colleagues who work to map out the distribution of galaxies like our own in the universe, have mapped one part in $10^5$, one part in a hundred thousand of the visible universe (the part that we can see).

How big is that? Well, that's about the fraction of the surface of the earth covered by the state of Rhode Island. Now in Massachusetts everybody knows where Rhode Island is, but even the people who actually live in Rhode Island probably wouldn't want to make generalizations about the appearance of the surface of the earth from a map of Rhode Island. One of the things that you've probably noticed already from the lectures that everyone has given is that there is no shortage of chutzpah in cosmology. And I have my share. So from this very small map, from this very small fraction of the universe, I will make wild generalizations. You should believe every word I say.

I'm going to tell the story in three parts. I'm going to begin by talking about making maps. Human beings have been making maps for a long, long time The story of our exploration of the universe is a story of our expanding horizons. Essentially the story of mapping the earth is a story of the bigger and bigger universe that we perceive. And as I tell you about the history of map making, I'll try to describe some of the things that contribute or enable this to be the age in which we can map the universe. Then I'd like to try to show you what you do if you want to make a map of something so unimaginably vast that you can only map a very small fraction and you want to get a result in your lifetime. I will show you how we design a strategy to answer questions about the structure of the universe. And then, finally, I'll show you the results.

Let me start by telling you something about the history of map making. A map of ancient Babylon, made in 500 B.C., appears rather primitive to us but it shows the Biblical cosmology and it contains the model of the universe of the ancient Babylonians. The Babylonians were not sailors. They knew algebra but they didn't know geometry. They didn't see ships disappearing over the horizon so they had every reason to think that the earth is flat. After all, if you only see a small piece of it, it does appear to be that. Just as Willie Fowler showed you, if you look at a small piece of the universe, you can think that's flat too. I didn't mean that as a slight, by the way, that's a real theorem in general relativity, that if you look at a small piece you can use Newtonian physics. That's another way of saying that locally the universe is flat.

In every direction the Babylonians looked they found water; they thought that their land was surrounded by water. One of the serious problems for the Babylonians were the great floods. They wanted to explain why these waters poured down on them and where they came from. Obviously there had to be something that held them up. And when this ceiling broke or opened up, the waters would pour down. In other words, there was a firmament or ceiling to hold the water up. That's how the word "firmament" originated. On this solid sky, the fixed stars were painted; it had little doors for the planets come in and out. This firmament was held up by seven mountains—the seven mountains of Biblical fame. This model of the universe appears rather silly to us now. This model covers about the same fraction of the earth that our maps cover of the visible universe!

Well, the art of map making took a long time to advance even though we can actually walk around on the earth. It's accessible to us. A map in Hereford Cathedral (England) dates from the 12th Century. And this map bears a remarkable resemblance to its ancient Babylonian cousin. The earth is still flat and is surrounded by water. The Hereford map is still more a representation of myth than it is of the physical structure that we call the earth. The map is distorted to place Jerusalem at the center of the map; Tensalem was the philosophical center for that era. The frontier had expanded to England. The map contains lots of instructions here about how not to get eaten by monsters in various parts of the world and so on. The map is a guide to survival, but not a very useful one.

What spurred people to make maps that really represented the earth? Commercial interests. During the Renaissance, there was an enormous surge in the art of map making partly driven by desire to travel around and trade. People wanted to survive the trip! Commercial and what you might call fine scientific interests came together. This coincidence of interests is often one of the reasons that science advances. For example, in astronomy today we have large infrared detectors. These detectors were developed primarily by the military; astronomy has profited greatly. Astronomy has also profited greatly from the development of technologies for a remote sensing.

The maps of the 15th century Renaissance were actually copies of maps made much, much earlier by Ptolemy. Most of us learn about Ptolemy as a man who got things wrong. He was sort of an idiot who thought that the universe was earth-centered. But actually he was a pretty smart guy. He recognized the usefulness of maps as scaled representations of physical systems. This idea is quite a profound idea which underlies much of science.

Today we make maps of all kinds of physical systems. We make maps of the arrangement of atoms in the DNA molecule. We understand the behavior of solids by relating the structure of materials to their properties. We try to understand the universe by making a map of it. Before we can understand a system we have to be able to describe it. To describe a system, we make a scaled representation of the physical system.

The copies of Ptolemy's maps—well-known features are readily identifiable—the Mediterranean, Spain, the north coast of Africa, the Arabian peninsula. But the maps are still distorted. People could measure coordinates north/south by observing at the fixed stars. But that couldn't measure longitude very well, they couldn't measure the coordinates east to west. Why? Think about what happens when you get on a plane to travel from the east coast to the west coast. You've got to reset your watch. To measure longitude you have to have good clocks. The reason that people couldn't measure longitude was that they didn't have very good clocks. Shipping companies began to sponsor competitions to reward clockmakers, not because they were so interested in clocks themselves, but because they were interested in maps. Once people began making clocks, they could make better maps. Here is an example of the development of technology which was supported by commercial interest helping science (map-making) along.

Very often in science fields are enabled by advances in technology. Certainly mapping the universe is one such area. In the 1920s Hubble first discovered the universal expansion. When he wanted to measure the redshift of a very nearby galaxy, a galaxy maybe one tenth as far away as the galaxies in our survey, he used the 100 inch telescope. He had to expose all night to measure a single redshift. Today, with solid state detector technology, we can measure a redshift for the same galaxies with a 60 inch telescope in less than five minutes. This change in technology enables us to map the universe. Without these advances it would be a hopeless task.

There's yet another point about the relationship between maps and science which we can see as we look to more and more recent maps. Maps made in the early 17th century still have a few important features missing. Antarctica and Australia are absent. North America generally has a few problems. But things were getting better and better. In maps of this period you can already see something which gave people an idea. South America appears to fit into Africa as though they're two pieces of a picture puzzle. By this time a few people had already begun to get the idea that maps were useful not just for telling you what a system is like today, but for telling you what a system was like—how it got to be the way it is. There were early speculations that the continents, the shapes of the continents of the earth, were not always the way we see them today. Perhaps they were once together in a single giant land mass.

The person who really pushed this idea came later—a man named Wegener. In the early part of the 20th century, he suggested in a very clear way the theory of continental drift. He found evidence for his suggestion by looking in the fossil record on coasts he thought were once adjoining. People ridiculed Wegener. They didn't take him very seriously. It took a long time before his suggestion became a part of what we consider the general core of knowledge of science. It was only in the 1960s that a French group took a small vehicle down to the bottom of the MidAtlantic rift. They were able to measure the rate of spreading of the ocean floor directly. Today, although the mechanism is far from completely understood, the model of plate tectonics or the model of continental drift is widely accepted as an explanation of the way the structure of the earth originated.

Good maps can make good science. It's pretty hard to understand a system if you don't know what it looks like! We have a lot of theories about what the universe is like, but until recently, we've actually had very little data. We've had no maps. When I was a student in the 1970s, there were many things that were not known about cosmology. They were not known because nobody had ever looked! Then some idiots like us went out and looked. It turned out we didn't know as much as we thought we knew. This kind of surprise is one of the exciting things in science.

One of the points of this story of mapping the earth is that in spite of our ability to walk around on the earth, it took us a very long time to understand, to map out the structure of the earth. There are still parts of the earth which are still not well mapped. We're still learning things from remote sensing about the map of the earth. Certainly we still don't understand its history.

The universe presents a very different challenge because we can only admire the universe from afar. We can never hope to travel around in it. And it is vast. It's so vast that in many, many scientists' lifetimes we will not map the universe in anything approaching the detail that we've mapped the earth. So what do you do when you have such an enormous system? How do you manage to find out anything at all?

The most important part of science is asking the right question. You've got to ask a question that you can answer. It has to be a question where the answer might be interesting. You can ask lots of questions which are not interesting. Nobody really cares much about the answer like why is my blouse so wrinkled now. I mean who cares? You can also ask questions that science can't answer at a particular time. What came before the Big Bang? What preceded inflation? These questions are examples—they are currently not really questions which are addressable in science. Of course, the questions we can answer change as we understand more. On the other hand, questions about how galaxies, like the Milky Way, are distributed on large scales are addressable with today's technology. The question has to be more precise. What about this distribution of galaxies? What exactly do we want to know about it?

Let me now pose the question. To pose the question, I have to tell you a little bit of history. It's been known for a long time that galaxies, like our own Milky Way, are clustered together. For a very long time these clusters which are a few million light years across were thought to be the largest "lumps" in the universe. Before we began our survey, you could look in textbooks and find the following statement: If you take pieces of the universe a hundred million light years on a side, all of the pieces should look more or less the same. But how big a piece of the universe do you really have to take before they all look the same? Obviously, it must be larger than the size of this room, the size of the earth, or even the size of the Milky Way.

People thought a hundred million light years was big enough. Not so—we now know that this idea was incorrect. And we really don't know how big is big enough. The only observation that tells us that there is a region big enough is the amazingly smooth microwave background Phil Morrison talked about.

In 1981 an observation first made people realize that there might be big sateens in the universe—that the universe was so uniform, so homogeneous on a scale of hundred million light years. Bob Kirshner, Gus Oemler, Steve Shectman and Paul Schechter did a survey to measure how many galaxies there are in the universe in every unit volume. They looked in one direction and measured how far away galaxies were in that direction. They they looked in another direction and they measured how far away galaxies were in that direction. They found that in three separated directions there were galaxies at some distance and then there were no galaxies and then there were some more. They found a region a hundred-fifty million light years across where there are essentially no bright galaxies. They christened this dark region the void in Boötes. Voids now commonly refer to huge empty regions where there are essentially no galaxies.

Well, people like John Huchra and myself, who were educated in the places where everybody knew the answer, figured that this result must be incorrect. There weren't supposed to be any structures of that size. We thought that there must be something wrong with these observations. Perhaps the survey was too sparse. Perhaps it was the only region in the universe that was so empty. After all there weren't supposed to be structures of this size! With this prejudice, it took us four years to realize that we actually had an instrument at our disposal to test whether the voids were common. At the end of those four years, we finally understood what the question was! The question was: Are there patterns in the universe, on a scale of a hundred million light years or a hundred-fifty million light years? Are they common? If so, what are they like?

Now, there are a lot of limitations to the way we could explore the universe. We had a sixty inch telescope, very small by today's standards, but big enough to map the nearby universe. A telescope big enough to map the nearest 2% of the visible universe. We had a telescope, but we were limited in the number of red shifts we could measure in a reasonable time. In a spring, for example, we could measure perhaps a thousand red shifts.

What fraction of the universe is that? Well, it's less than one part in $10^5$. We want to map one part in $10^5$ or less and we want to answer a pretty grand question. Does the universe have continents and oceans? Does it have big patterns; and if so, what are they like?

Let me explain the strategy we followed by making an analogy with exploring the surface of the earth. Suppose that you are approaching a cloud-enshrouded earth on a spaceship. You're going to get to see one $10^{-5}$ of the surface of the earth. You have to figure out whether it has continents and oceans and if so, how big are these features. What would you do? Let's say you observe a random patch of the earth which covers $10^{-5}$ of its surface. Most of the time it would land in the ocean: it wouldn't answer your question. Here's another random patch which ought to be familiar to you here but is a lot less familiar on the east coast. This picture shows your local region of the universe—it certainly doesn't answer the question. It says the earth is pretty boring, flat. The earth's surface is one kind of stuff, no structure. A randomly chosen small patch doesn't do the trick.

There is a simple strategy which will work. Well, this is a map of the earth and Rhode Island, somewhere here. If I take a great circle around the earth, in almost any orientation, I will pass through continents and oceans. I can make this strip arbitrarily thin. This strip can be so thin that it covers only one part in ten to the fifth of the earth. What do I have in a strip around the earth? I have ocean, continent, ocean, continent, ocean, continent. I learn that the earth has two kinds of structures and that they're both big. That's a lot to learn for such a little!

There's something wrong with that argument about strips around the earth. There are some strips which don't pass through continents and oceans because the oceans are connected and the continents aren't. I could choose a strip which turns out to be ocean only, but those strips are rare. This problem brings up a point which we worry about in astronomy called selection bias. Sometimes we get a sample that isn't typical of the universe. We're going to assume that the universe doesn't know that we're trying to observe it from a mountain in Tucson. We assume the universe couldn't care less and we assume that the samples we take are typical.

The universe is a little more complicated than the two-dimensional surface of the earth because it's a three-dimensional place. The analogy to a strip on the surface of the earth is going to be a slice in three-dimensional space. To sample the universe I first look up at the sky. We look at galaxies in a strip across the sky. We have to measure how far away they are. We take a slice of the universe and figure out the distribution of galaxies. We know the positions of galaxies on the sky and we then measure how far away they are.

Let me tell you a little bit about galaxies. Philip Morrison has already talked about galaxies, but let me do it too. Our own galaxy, the Milky Way, is a spiral. Galaxies like ours have a bulge which contains an older population of stars and a disk where you find younger stars like the sun. It is far from the center of our galaxy. The Milky Way is tens of thousands of light years across. The masses of galaxies are very uncertain, but they are about a trillion solar masses, a trillion times the mass of the sun. There's one solar mass in a galaxy for every dollar in the federal deficit. At least for now.

In nearby galaxies you can see the beautiful spiral structure. Galaxies don't have well-defined edges. One of the great mysteries, one of the great challenges to those of us who work in this field is that 90%, or perhaps even 99%, of the matter in the universe is dark. Most of the matter doesn't give out any light. It's hard to know, as you learned this morning, what the matter is or where it is. When we map the universe, we mean that we make a map of the distribution of objects that give out light and those objects are galaxies—giant collections of stars, gas, dust and dark junk.

It was known by Hubble that galaxies are not randomly distributed in the universe; they cluster together. Clusters of galaxies contain hundreds to thousands of galaxies like our own Milky Way. Sometimes the galaxies are so close together that they interact. In the visible universe, there are some ten billion galaxies similar to the Milky Way.

The first thing we do to map the universe is to take pictures of the sky. From these pictures we learn the latitude and longitude, the positions of the galaxies, on the sky.

Next we take these pictures to our observatory at Mount Hopkins, south of Tucson. On Mt. Hopkins we have a 1/2 meter telescope which is the work horse for the mapping the universe project. Every night, every dark night when there's no moon and it's not raining in Tucson, there are technicians who point this telescope at one galaxy after another to map the universe. They measure how far away the galaxies are one by one. It takes 15 minutes to a half an hour for each galaxy. With this telescope we measure about 1500 red shifts a year. We've measured something like 13,000 redshifts with this telescope, about 1/3 of the known red shifts of galaxies.

Our 1.5 meter telescope is not a very sophisticated telescope which is very fortunate for us because nobody else wants any time on it. The telescope was very cheaply built. It has a spherical mirror. Nobody in in their right mind would build a telescope that way today. But the telescope was cheap. Because it has a spherical mirror, it can't image worth anything. The only thing it's good for is collecting light to take spectra. That's what we use it for. This telescope is the work horse for the survey.

How do you find out how far away the galaxies are? You know their positions on the sky. You know their latitude and longitude. Now you want to know how far away they are. You find the distance by using Hubble's Law for the expansion of the universe. Hubble discovered that when we look out at the universe we see an apparent expansion. We see that the galaxies recede from us with velocities which are just proportional to their distance. Space is expanding. This concept seems rather mysterious at first glance, but in fact it's a very simple kind of expansion. A simple one-dimensional demonstration is a piece of elastic with stars sewn on it at regular intervals. These stars represent galaxies like the Milky Way. For a simple example, start with them uniformly separated. Stretching the elastic is exactly like the Hubble Law. From any galaxy (star on the elastic), we see the same kind of expansion. What do we see? If we sit on a particular galaxy, its heaviest neighbor moves away from it with some velocity v. The second nearest has to move twice as far so it's going twice as fast and so on. One very interesting thing about this expansion is that you notice that if the stars are uniformly spaced when we start, they're uniformly spaced when we quit.

The expansion by itself doesn't change the geometric arrangement. It only changes the scale. The Hubble expansion is a simple stretching. The velocity is just proportional to the distance; this is the Hubble expansion.

For galaxies we actually measure velocities, not distances. That's how we find out how far away galaxies are. Now how do we measure these velocities? Well we measure them by looking at the spectra of galaxies. We point the telescope at one galaxy after another. We spread the light from each galaxy out into its colors. In the spectrum of galaxies there's a pattern of lines. This pattern reflects the properties of the elements and ions in the material in the galaxy. We can recognize these standard spectral features. The strange thing about the features is they're shifted to the red; they're shifted to a longer wavelength than where we would see them in the laboratory. You see an example of line emission if you walk on the streets and you look at sodium lamps. They're yellow. They're yellow because of the lines of transitions in the element sodium. We see sodium lines in galaxies, but they're redder than the lines we see from the street lights.

That's because the galaxies are going away from us. It's very similar to an effect you've all experienced, the Doppler effect. When you stand on a street corner and you hear a fire engine go by you, you hear the pitch drop. This is the Doppler effect: you have a shift to longer wavelength. Another way to think about this, which is actually more correct, is that the space of the universe stretches as it expands. Photons travel in this stretching space. As the universe stretches, so does the wavelength of the photons. This effect gives you a redshift. The features are shifted to the red.

Galaxies typical of our survey are too far away to show much structure in a picture. When we spread the light out into its colors, we can see here, at a red wavelength—a line of hydrogen. Hydrogen is the most common element in the universe. We see a line which tells us that this galaxy's got hydrogen. If we look at another galaxy, it might appear smaller, but that doesn't necessarily mean it's farther away because galaxies vary in size. If I measure the redshift—I spread the light into its colors—the spike—the line of hydrogen—appears at a longer wavelength. Why? This galaxy's farther away; it's moving away faster. If we look at galaxy spectra one after another, we have a set of redshifts. Then I can map. I have latitude, longitude and distance. I can make a three-dimensional map.

The first thing we need to do to map the universe is to find the positions of galaxies. Fortunately this work did not have to be done by us for this particular project. It was done by a man named Zwicky and his collaborators in the 1960s. They looked by eye at a thousand plates of the sky and they plotted the positions of 30,000 galaxies. Today you would do this by machine but they didn't have machines capable of the task. Today people do it for millions of galaxies by machine.

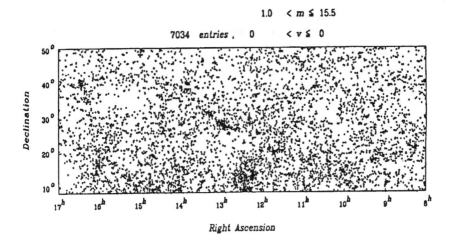

Figure 1:  Distribution of galaxies on the Northern sky.

Figure 1 is a map of the positions of 7,000 of those nearby galaxies on the Northern sky. Each little cross represents a galaxy like the Milky Way. These galaxies are nearer to us than about half a billion light years. The coordinates are the latitude on the vertical axis and the longitude on the horizontal one. At 13h and 30°, this is a dense knot of galaxies called the coma cluster. It's nearly over the North Pole of our galaxy. Now you might ask why doesn't the map cover the full 24 hours? Because we can't easily look through the plane of our own galaxy; there's lots of dust and junk in the plane of our galaxy, so we can't observe through the plane at optical wavelengths. We have to look up toward the Pole because light is absorbed by dust in the plane.

You can see that the distribution of galaxies on the sky is not random. There is clustering. You can see that there is structure here, but it's nothing to write home about. If you cover the figure and try to reconstruct it without looking, you probably would have trouble. There's no distinct pattern. I wouldn't be here talking to you if this were the end of the story.

We had to decide what to observe. We didn't expect any remarkable results. We followed the analogy to exploring the surface of the earth. If we want to look for large structures, we should measure redshifts for galaxies in a strip across the sky. And we suggested this project as a thesis for Valerie de Lapparent, who was then a graduate student working under my supervision. In a strip we could measure about a thousand redshifts. And what strip did we choose? We chose a strip that's overhead in Tucson. It crosses the Galactic pole. The choice minimizes what we call selection effects. We assumed, of course, that the universe doesn't care where Tucson is. Our strip should not be one of those special strips that has only ocean.

We were so sure that there wasn't going to be an exciting result that we didn't even encourage Valerie to plot up her data until they were complete. John and I one evening were sitting in his office talking about Valerie's thesis. John asked me, "What in the world is Valerie going to do for her thesis?" "Well" I said, "Well you know it's a thousand new redshifts. We'll find something. There are standard statistics that she can measure. Don't worry. It's just a thesis."

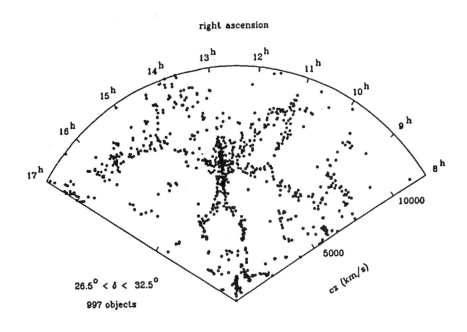

Figure 2: Distribution of galaxies in a first slice of the universe.

Then one day we saw the data. Figure 2 shows it. This pattern is one you can remember. Everyone calls it "the stick man." The earth is at the vertex. The outer limit of the survey is at a distance of some 500 million light years. Each of these points is a galaxy approximately equivalent to the Milky Way.

There's a pattern, an easily described pattern. There are big, empty regions surrounded by very sharp structures.

It was an extraordinary experience to see this pattern. The three of us had very different reactions. John Huchra says that he worried that we weren't supposed to see something like this and therefore there must be something amiss in the data. I felt a thrill because I worked with John for 15 years and I trust him absolutely. I was sure that there was nothing wrong with the data, that the stick man was something we had to explain. All of us felt this extraordinary thrill. Why should we be the first three people to ever see this pattern? It was an extraordinary feeling. Here is territory that no human being had seen before. Here we were, the three who saw the pattern for the first time. It's one of those thrills that makes life worth living. It is a remarkable experience to have in a scientific career. If you have it once, you're really blessed.

Well what did we see? We saw vast empty regions. 150 light years across just like the void in Boötes. I went home that evening and wondered, "If this is a typical slice of the universe, what does it say about the geometry of the universe? What kind of structure could you slice through and have every slice look like this?"

There are many structures in nature which do this. These are structures which have sheets and holes. Things like soap bubbles. If you take a slice through the bubbles in your kitchen sink and the thickness of the slice is small compared to those bubbles, you'll see the soap film surrounding the empty interiors of the bubbles. If you look at the edge of a sponge it might look like this. Now sponges and bubbles are different topologically but basically the idea is that you have thin sheets surrounding empty regions.

We announced this result in January of 1986. As soon as we did that, we got quite a bit of help from the media in interpreting it. A newspaper in Green Valley ran the headline "Lawrence Welk Universe Theory." Green Valley is a retirement community near the base of Mount Hopkins where the observations are made. My favorite is the end of this article which says, "Imagine Lawrence Welk playing God. Blowing those bubbles while the band plays Stardust. Who knows, maybe the universe is just one big puff piece."

We were instantly famous and instantly a subject of controversy. The burden on us was to show that the suggestion we had made about the structure and the commonness of the holes and thin sheets really had some merit. Valerie, needless to say, got a degree and a permanent job pretty quickly. John and I were left to slog away to figure out what was going on. The next spring we did the obvious thing; we wrapped the distribution of galaxies in the adjacent slice of the universe to the north and found the pattern in Figure 3. Once again you see big holes and very thin structures surrounding them.

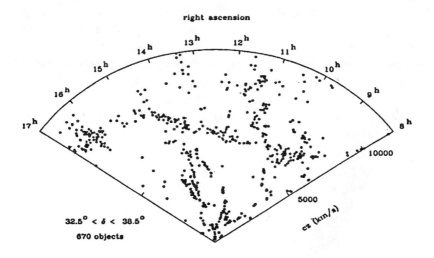

Figure 3: Distribution of galaxies in the slice adjacent (to the north) of the slice in Figure 2. Note the remarkable correspondence of the structure.

The slice in Figure 3 is adjacent to the first one. Did we celebrate or were we ready to go out and hang ourselves? Does this pattern match or not? Well the match is extraordinary.

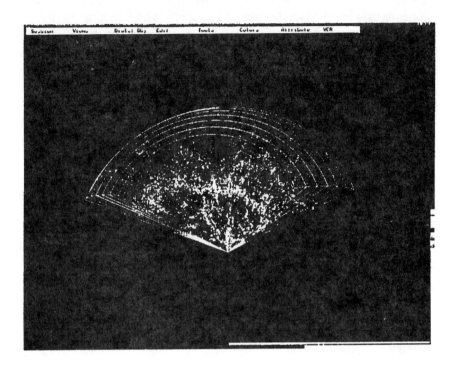

Figure 4: Four slices of the universe: Three are adjacent. There is a gap of two slices, and then a fourth. Note that the main structures are the ones already visible in Figures 2 and 3. The pattern here is as big as the survey.

Figure 4 shows you the data in four slices. Let me say a little about the challenges these data pose. The basic problem which you've heard many times in this series of lectures is that we have two pictures of the universe. We have a picture of the universe when it had an age of a hundred thousand years; then it was very smooth. And we have a picture of the universe when it has an age of 10 or 20 billion years (today) and we see enormous structures.

How do we get from one to another? The simplest thing is to move the matter out of the big voids. The shortest distance you can move them is the radius of the voids. My picture is that the voids are the clue to the evolution of this structure. But the simplest models don't allow you to make voids this large and structures this thin. There have been a variety of suggestions to help out. Perhaps there were explosions in the early universe. These explosions may have driven the matter into shells which fragmented to form galaxies. These explosions would leave an imprint on the microwave background (or people speculate that it would) and that's no good because we don't see it.

Among widely made suggestions is that the dark matter in the universe is fairly uniformly distributed. The voids are dark but not empty. Galaxies form only where the density slightly exceeds a threshold. For an analogy, think about looking out at the ocean. You look at the choppy sea. There's foam on the crests of the waves. Let the foam represent the galaxies, and just forget the sea. The galaxies are just foam on the crests of waves on a choppy sea. You just don't see most of the sea which is a very smooth, dark material. There are many problems with this model. The supposed dark matter particles have never been detected in lab. Nobody understands why it is that galaxies should form just when the density threshold is just right. It's not clear that the models can produce structure which match the data.

We have a puzzle. We have two pictures. We don't really understand how the furniture, how these patterns in this universe, were made. One of the great things about the next 10 or 20 years is that we're constructing large telescopes on the ground which will enable us to look into the universe to many times the depth of this survey—maybe 20 times, 30 times the depth of this survey. If we can't figure out by pure thought how these patterns got to be here, we'll see it some day.

A video which displays the data in 3-dimensions, "Where the Galaxies Are," is available from the Astronomical Society for the Pacific (390 Ashton Avenue, San Francisco, CA 94112 - Tel. 415-337-1100).

# Questions and Comments

**Fuller:** Let me remind you if you have questions, again you can send them to the aisle for the ushers. We are going to start with comments and questions from the panel. Professor Ferris.

**Ferris:** I would like to ask a question, I'm sure many members of the audience are wondering why, with such important work, were there not bigger telescopes and why is there not a bigger program in this field at this time?

**Geller:** Well there will be. I think one of the things that happened is that no one really expected the results we found. And so when this result, when this pattern showed up in '86 it took awhile for people to recognize it. These patterns make it clear what some of the questions really are. Now there are many many groups which are mounting large projects. For awhile we had very little competition. That's bad and good at the same time. But there are a number of groups working now. There's a collaboration among Princeton, Chicago, Fermi lab. They're planning to build a 2 and a 1/2 meter telescope dedicated to mapping the universe and they'll map the universe out to a redshift of .2 or so.

We have convinced Harvard University and Cambridge University in England to try to raise money to build a 4 meter telescope in the southern hemisphere. The idea is that this telescope would be dedicated to large projects. The telescope would have a large field of view so that you could measure many, many red shifts at a time using fiber optics. We decided on a 4 meter because we think that the most interesting questions by the time this telescope will be built is to try to address the evolution of the structure. When you look out in space you look back in time. If you have a big enough telescope you look out far enough in the past to figure out what happened. The Smithsonian operates a 4 1/2 meter telescope on Mount Hopkins. This telescope is going to be replaced with a single 6 1/2 meter light weight mirror. This telescope will have a 1 degree field of view and we will have an instrument with a fast robot with 300 fibers which can be positioned in a minute or so. We'll be able to measure galaxies at a redshift of .2, 300 galaxies at a redshift of .2 in half an hour.

**Ferris:** Bigger programs.

**Geller:** Yes.

**Fuller:** Professor Fowler.

**Fowler:** Now I'm puzzled. This structure that you see. How long is it going to last? Will it eventually be all smoothed out or will there be more and more structures as the future of the universe?

**Geller:** Well it depends on Omega in part.

**Fowler:** But Omega's one now, you know that.

**Geller:** It is not.

**Fowler:** All right. All right. I'll grant you that.

**Geller:** No you won't.

**Fowler:** No, I'll grant you that it depends on Omega.

**Geller:** Oh, I thought that was too easy. It depends on the evolution. We've made some heuristic models where if you start out with holes in the universe, they tend to grow and they sweep out matter into thin shells. The shells grow. The bubbles get bigger and bigger. The walls break and you see bigger and bigger bubbles with thin shells around them if Omega's one. If Omega's low, the structure essentially freezes in and just stretches. You don't see the clusters grow. One of the problems with these models is that the epic over which you see structures like this is not that long. I don't know that anyone knows the answer. In the simple models we've made, you don't have a very long time over which the structure really looks like the data. The models don't yet match the scale of the structures. It's one of the most important questions.

**Fuller:** Professor Fowler.

**Fowler:** Well the contrary question. How long, it's only been 10 or 12 or 13 billion years . . .

**Geller:** Right.

**Fowler:** And this structure has developed in that time.

**Geller:** Yes, and that's one of the difficult things. One of the difficult things is producing large enough structures in such a short time.

**Fowler:** Yeah, Yeah.

**Geller:** If the universe were older, it would be a lot easier to make the structure. That's one of the exciting things about trying to make maps like this at high redshift. Then you would have another point on this nice curve that Phil drew. You would know something about the scale of the structure in more than one epic. I think in the next 5 to 10 years we'll know rather well what the characteristics of the structure are nearby. There are some indications already that there are others of these big structures out to redshift of a couple of tenths. They're common, but the question is when do you see evolution and at what rate. And fortunately I think we'll be able to measure the evolution directly.

**Fuller:** Professor McMullin.

**McMullin:** Yes, you said in passing I think partly quoting Phil here, that at some level of size you would expect to return to homogeneity again. That the theoretical model would suggest that at some scale one would return to some kind of smoothness. Is that in fact a conclusion of the model?

**Geller:** The microwave background tells you that the early universe was highly isotopic and homogeneous. If the interpretation of that is correct then there must be some scale at which you do see homogeneity. It could be that people who argue that Omega is one in elementary particles are correct and that the universe is actually very smooth. They argue that the light emitting matter, all this beautiful structure that we see, that's just the flotsam and jetsam. It's all irrelevant to the evolution of the universe. It just happens to be what we see. So we're observing the garbage and the main stuff is all smooth and boring.

**McMullin:** I wonder, I wonder if I could just ask Phil to possibly say something about that. But as part of the garbage I guess I object. But..

**Geller:** I can't tell if he's laughing or crying.

**Fuller:** I can't tell either.

**McMullin:** If I could pursue that one stage further. Your work and that work of others has already shown an enormous degree of inhomogeneity. Now . . .

**Geller:** These structures are still small you know. It's only 5000 kilometers per second or 150 million light years opposed to the 20 billion light year radius of the universe. It is pretty small still.

**McMullin:** That's still . . .

**Geller:** It's a percent.

**McMullin:** I know, but that's still inhomogeneity and if that can be explained I wonder why you have to insist that at some scale you have to return to homogeneity?

**Geller:** Well, Phil, I think . . .

**Morrison:** Well there's lots, Margaret already explained to us quite correctly that her colleagues have enormous brass and they don't mind concluding big things from small conclusions. And they're also terribly difficult to satisfy. They don't let well enough alone; they're always going to ask questions about this wonderful work and the beautiful images which I think are the best graphics ever presented on this subject by far. You live in a highly heterogeneous region, Professor McMullin, you always have. Down below you, you step on something that has a density one. Whenever you walk around, you walk through something that is density ten to the -3. You go a hundred miles up and it's density ten to the -9 or 10. Go a little further out, though the tiny astromica is ten to the -24. And then bang you hit a star or a cluster and the thing will go up. And this is the way of the world. That is the nature of gravitation.

Quantitatively certainly we cannot explain these structures, I think that's quite clear. We're going to learn a great deal when we do explain them. But it doesn't seem to me it's a grand mystery. It seems to me it's rather tantalizing evidence how the simple laws of physics can give results which we already should know. And we always find out, that it is really very simple looked at the right way. We haven't looked at the right way until

Margaret and John looked at it the right way and then they see something we didn't expect.

But if you average over this region, give it a number, it's some number, and I bet that the average of the next region of the same size is a rather similar number. That's what she would say.

**Geller:** No, not this, not yet.

**McMullin:** Well have you done it?

**Geller:** Yeah, not yet.

**McMullin:** How many have you done around the sphere?

**Geller:** We have three. We have three.

**McMullin:** Now are we near a sphere of a 150 million?

**Geller:** We have three.

**McMullin:** . . . 400 million

**Geller:** . . . of this size. The regions are not yet big enough because the uncertainty of the mean density goes as the square root of the number of the biggest voids. There's only a couple of voids in regions of this size, the regions are not big enough.

**McMullin:** So you have to find the sphere? Yeah.

**Geller:** Right. There is a new survey by Kirshner and his colleagues.

**McMullin:** Yes, yes. Of course that's step one.

**Geller:** Yeah.

**McMullin:** But step two is of course you're not seeing everything as we know there may be dark matter. We don't know.

**Geller:** Right.

**McMullin:** But I want to ask a more particular question. What is the amount of luminous matter, either in mass or in light terms fractionally, that you miss. You must have some limit to the survey.

**Geller:** Oh, there's a magnitude limit.

**McMullin:** A magnitude limit or a mass limit?

**Geller:** There's a magnitude limit which is 15.5. When we do a survey, the galaxies don't come with little tags telling us how far away they are. We only know their apparent brightness. We observe all the galaxies which are brighter than a certain amount. When you look at the outer boundaries of the survey you see very few galaxies. That's because you see only the intrinsically brightest ones. I think that Phil is asking, how far down the luminosity function are we seeing? Galaxies have a wide range in luminosity. In the regions where the survey is densely populated, we see about half.

**Fuller:** We have a couple of questions from the audience. Question concerning the voids. In particular, what drives them and maybe this has been answered in the comments you've made. In particular, how did they come about?

**Geller:** How do you start it and then how does it grow? Gravity behaves in a funny way when it comes to voids. If you start with a lump, a positive lump, a denser region, denser regions like to get bigger. But voids, low density regions also like to get bigger! Gravity behaves in this very funny way. If you start a low density region then it'll act like a lower density region embedded in a universe which is on average at higher density. The low density region wants to expand faster than its surroundings. It piles up matter on the edges just the way you see. How do you start it going? Well we don't know. A week ago or so Phil and I had a discussion. We met at a party and he made some suggestions of how to start these. I think there are many people, including Phil, who are trying to figure out how you get it

# Long Ago and Far Away: Cosmology as Extrapolation

Ernan McMullin

There was a front-page story recently in the *New York Times* about a cosmologist. Nothing strange about that, you might say: cosmologists are in the news all the time nowadays. But this cosmologist lived four centuries ago, despised mathematics, and based his belief in the infinity of the universe and the plurality of inhabited worlds on the infinity and overflowing generosity of God. Not your standard cosmologist, even in those far-off days. But it was not for his cosmology that Giordano Bruno was in the news once again. An English historian has concluded that Henry Fagot, the mysterious spy who in the 1580's betrayed a conspiracy on the part of English Catholics to overthrow Elizabeth I in favor of Mary, Queen of Scots, was really Bruno, who at that time was a resident in London at the home of the French Ambassador.[1] Secret letters from Fagot led to the arrest, torture, and execution of the conspirators. Circumstantial evidence links the writer of the letters with the French Embassy. Circumstantial evidence also suggests that Bruno may have done a little "intelligence" work on the part of his hosts at the Embassy. Could he have been a double agent? He was assuredly in desperate financial straits at the time. The author of this new hypothesis, John Bossy, in a book about to appear soon from Yale University Press, argues that all signs point to Bruno as the "mole" in the Embassy who betrayed the secret understanding between his host and the conspirators.

Good Le Carré, you say, and indeed the story has many of the elements of an imaginative thriller. But did it really happen that way? Was one of the distant forerunners of many-universe cosmology a double agent for France and England? Barring the discovery of new evidence, we may suspect, we may debate, but will never have anything more than a suspicion of what might have happened. The few lines still accessible to us that lead back to that distant pattern of human action are too uncertain, too hard to read.

Historians often find themselves unable, with any degree of assurance, to reconstruct events from the past, even events that occurred only a short while ago. The events have not left sufficient traces behind to enable one to infer what actually happened and why. This is the main source of that combination of frustration and fascination that writers of history know so well. How is it possible, then, that the special breed of historian we call cosmologists can lay claim to reconstructing in fair detail the sequence of events at an epoch more than a hundred million centuries in the past? What kind of special magic do they have that their historian-colleagues who concern themselves with political intrigue in late sixteenth-century Europe seem to lack? And how far can we trust that magic? These are the questions that the title: "Cosmic evolution: How do we know?" encouraged us to ask.

I am going to take a rather indirect route in an attempt to answer these questions. My concern will be epistemic. I want to ask first about the quality of knowledge that cosmology afforded in its distant past. What were the criteria that the earliest cosmologists relied on in proposing speculative extrapolations into the distant and mysterious realm of the planets, extrapolations as daring in their way as those of the big bang theorists of today? Next, how did the proponents of the "new science" of the seventeenth century propose to secure the precarious-seeming inferences from effect to remote cause that they were constantly drawing? Then, what in outline were the steps that led over the course of the past half-century to the acceptance of physics-based cosmology as a science? At that point, we should be in a better position to face contemporary cosmologists with the crucial question: how do you discriminate the credible from the merely ingenious in all this talk of bangs, crunches, and parallel universes?

## 1.  *Mathematical versus physical astronomy*

The most distinctive contributions to the early development of astronomy came, I suppose it would be generally conceded, from Babylonia and Greece. The astronomies of these two great civilizations were strikingly different in character but each, we can now recognize, bequeathed essential elements to those who would come later. The Babylonians from a very early period associated the phenomena of the skies, notably eclipses and first and last appearances of the planets and certain bright stars, with great events on earth: wars, plagues, droughts, attempts on the ruler's life, and so on. The celestial phenomena were seen as omens, as messages from the gods. An omen list, the *Enuma Anu Enlil,* associated specific appearances in the sky with equally specific occurrences of a large-scale sort in the kingdom. Those charged with interpreting the omens would send reports from the major cities of the kingdom of what the skies portended in the months to come. This gradually led to systematic observation of the heavens on their part, and the keeping of detailed records. Periodicities could then be estimated, making use of the relatively sophisticated arithmetical methods already available.

By the seventh century B.C., a complex mathematical-observational practice had begun to take shape; by the third century B.C. it had become an autonomous study of celestial phenomena, no longer tied simply to the reading of omens. The story is a fascinating one for many reasons, not least because it shows how a practice we can recognize as a precursor to the sciences of today could develop out of an activity of a quite different sort whose goal was surely *not* the systematic study of Nature.

How would the mathematical astronomers of Babylon have answered our theme-question: "how do you know?"? They might have said: "We have carefully observed lunar eclipses over many years, and have discovered a pattern which keeps repeating. Our mathematics gives us an economic summary of this pattern, and thus enables us to tell when the next eclipse will occur." The underlying assumption here is that the future will be like the past, so that the mathematical ordering found in celestial phenomena in the past can be extended into the future. No assumptions are made about the causal mechanisms involved. The mathematical formalism is an *instrument* serving the purposes of economic description and accurate prediction only.

This sort of *instrumental* knowledge is governed, clearly enough, by two rather different sorts of criteria. One is epistemic: the astronomer has to get the predictions right; if he doesn't, the formalism may have to be modified or the record-keeping may need to be improved. One can imagine a Babylonian magus, after a failed prediction of a lunar eclipse, puzzling over the records and asking himself whether errors had been made in recording past eclipses or whether the formalism itself needed to be altered in some way. The second criterion is pragmatic. Since the formalism is regarded merely as an instrument for use in description and prediction, the astronomer is likely to ask whether it is convenient, easy to use.[3] Though ease of calculation was obviously an important factor in the choice of method, modern scholars have marveled at the complexity of the mathematics underlying the ephemerides or almanacs found so frequently in the cuneiform tablets from the latter days of Babylon. Predicting lunar eclipses, it is worth emphasizing, was as esoteric a task in terms of *that* day as reconstructing the first moments of the big bang would be in terms of ours.

Despite the triumph of their mathematics, the Babylonians (so far as we can tell) never made use of the astronomy to construct a new *cosmology*. The interest of their astronomers seems to have been restricted to the kinds of events that appeared in the original omen-lists: eclipses, first and last sightings of planets, and the like. They did not chart the *paths* of the planets;

their formalisms always remained arithmetical. They did not discuss *why* the planets moved as they did; the original belief that the celestial configurations represented messages from the gods would have supported an unquestioning assumption that the planets moved as they did simply because the gods willed them to do so. It would seem that the astronomical knowledge of Babylonia remained at the instrumental level to be judged primarily by how well it described the limited set of phenomena from which it claimed to extrapolate the future.

And so we come to the Greeks, who developed a very different sort of knowledge of the skies, to be judged by different criteria and yielding (it may fairly be said) the first more or less "scientific" cosmologies.[4] The intent of Greek astronomy prior to Ptolemy was first and foremost to *explain* the celestial motions, and to use this explanation to construct a model of what the cosmos, the universe regarded as an ordered whole, might look like. There was less emphasis on exact observation and on record-keeping than there had been among the Babylonians, though some systemic observation was carried on in the later period. Instead of calculating moments of first appearance and the like, the Greek astronomers followed the apparent *paths* of the planets, making use of geometry rather than arithmetic. Plato is said to have posed astronomers the problem of accounting for the apparently irregular motions of the seven planets (the "wanderers") by means of combinations of circular motions. The circular motion was allowed a special explanatory status mainly because it was the only motion that returned on itself and could therefore be regarded as stable. Only such a motion could help to explain the everlasting movements of the heavenly bodies. "Saving the phenomena," the phrase later attributed to Plato, did not have the meaning for him it came to have afterwards for those who saw in the mathematical constructions of the astronomer merely calculating devices, furnishing instrumental knowledge only. The warrant for the circles was not so much that they exactly described or predicted as that they made intelligible *why* the planets moved as they did. But did they *really* explain the planetary motions?

Aristotle evidently did not think so. According to *his* theory of explanation, the most sophisticated one of the ancient world, to explain a natural motion requires one to specify what the mover is and how it operates, as well as indicating the order to which this motion contributes. Though things can move in circles, circles of themselves can't move anything. Some kind of *agency* is required; something, in short, must be carrying each planet along.

What could that be? The clue came from Eudoxus, an astronomer contemporary of Aristotle's, who had proposed an ingenious system of mathematical spheres to "save", that is, account for, the irregular motions of the planets. Each planet was given two basic circular motions, a daily one and a yearly one (corresponding, in our terms, to the two motions of the earth). The sun and moon were allotted one more circular motion each, the remaining planets two (to account for their occasional retrograde motions, when they seem for a time to move backwards in the sky). Each of those circular motions had its own characteristic speed, and its own axis of rotation. What simpler, then, than to think of Eudoxus' mathematical constructions as standing for *real* spheres, capable of mechanically interacting with one another, so that the composition of motions envisioned by Eudoxus could be brought about by contact action between nested spheres, each with its own axis and speed of rotation. Aristotle did not have much to say about the nature of the spheres themselves, but later generations would note that they would have to be transparent to allow starlight through, incapable of decay in order that the planetary motions would continue eternally, and different in character to any terrestrial substance since the natural motions of the spheres are circular, unlike the rectilinear natural motions found on earth.

All very plausible, once the original postulate of solid sphere be admitted. But why should it? What would Aristotle have answered if someone had asked him: but how do you *know*? One answer he could not have given, and which it would not have occurred to him to give, was the Babylonian one: that his model accounted precisely for the observed planetary positions. Though he added 29 more spheres to the 26 postulated by Eudoxus in order to make his system into a neatly interlocking quasi-mechanical one, he could not have made it yield exact predictions, since some of the parameters, as far as we can tell, remained indeterminate.[6] Qualitatively, his system *did* account in a general way for the main features of the planetary motions including the retrograde loops, and that is clearly all that Aristotle asked of it. Its real warrant lay elsewhere: the concentric system of spheres gave a plausible causal account of *why* the planets move as they do. And this account was plausible precisely because it satisfied the intuitive principle of contact action which underlay Aristotle's whole way of making sense of motion.[7] We shall call this a principle of natural intelligibility. How did he *know* the planets were carried on spheres? Because only this would make *sense* of the situation, even if the existence of the spheres could not be tested in any more direct way, by predictions of the form: "if there are solid spheres, then . . .".

How does all of this fit in with big bang cosmology? The status of Aristotle's spheres was, in some respects, not all that different from that of the many-universe models that have proliferated in the past decade. Perhaps the comparison is not fair to Aristotle since, in fact, the existence of the spheres was relatively easy to test. Aristotle himself had linked the brightness of heavenly bodies with their nearness, and it was an easy inference that if their brightness varied, their distance from earth also must vary. The planets, particularly Venus and Mars, clearly altered brightness in a regular way. How could this be accommodated in a system of concentric spheres where each planet travels on a circle with the earth as its center? Not at all, it would seem.

In the century after Aristotle, two new geometrical models were devised by Apollonius and Hipparchus to account for planetary positions. One was the eccentric, where the planet moves uniformly on a circle centered on a point at some distance from the earth; the other was the epicycle, where the planet rotates in a circle whose center itself rotates on another circle, the deferent circle, whose center is in turn the earth. One immediate advantage of these devices was that they might be made to explain the varying brightnesses of the planets, since each of them makes the distance of the planet from earth vary. But it was not until the Alexandrian astronomer, Claudius Ptolemy, who flourished in the early second century A D., that a system based on eccentrics and epicycles was worked out in detail. The interest for us of Ptolemy's great work, the *Almagest.* is that it gave an excellent mathematical depiction of planetary paths as viewed from earth, so good indeed that it remained the standard in this regard for fifteen hundred years. But it did not attempt to *explain* these motions; it seemed on the face of it difficult (if not impossible) to interpret the epicycle, and other even more complex motions that Ptolemy was forced to employ, in terms of solid spheres.

And so began a very strange period in the history of cosmology, one that would last until the time of Galileo, a period when there were, essentially, two rival astronomies quite incompatible with one another, each claiming certain virtues, but each displaying shortcomings from the perspective of the other.[8] To call them rivals might suggest debate and disagreement as to which of the two was to be preferred. But this was largely avoided by calling one of them "physical" and one "mathematical", and implying that their functions were different and that they should be judged by different criteria One cosmology purported to explain causally why the planets moved as they did; the other claimed to provide a mathematical scheme that would enable specific predictions to be made. But which could be relied on to tell us how things *really* are? Planets can't move both in concentric circles and on epicycles at the same time. Here there was a certain amount of disagreement, but the majority view was that the "physical" or causal picture given by the nested solid spheres conveyed how the cosmos *really* is. The mathematics of Ptolemy would then be no more than a practical calculational device, and thus not the basis for a cosmology proper.

But if the spheres were real, why did the cosmology based on them not predict better? Why could it not respond to the obvious anomaly of varying planetary brightnesses? There were no answers to these questions. And there was less apparent worry about this fact than one might have expected. Some, like the great Arabic natural philosopher, Averroes, found troubling the separation between the mathematical and the physical, the calculational and the causal, and urged efforts to devise a single astronomy that would bridge these dichotomies, one that would afford both causal understanding and accurate prediction at the same time. But, on the whole, most philosophers seemed resigned to the thought that a physical cosmology that did not quite save the phenomena might still be satisfactory, might still lay claim to *knowledge*. In short, the criterion of explanatory appeal was given *far* more epistemic weight than that of predictive accuracy. And explanatory appeal itself depended on what one took to be the permissible modes of bodily action. If action could be transferred *only* by contact, then Aristotle's system had enough in its support to fend off the critics. But there *was* one other way in which the principle of contact action could be satisfied.

In 1577, the great Danish astronomer, Tycho Brahe, calculated the elements of the orbit of a particularly bright comet and showed that it would have to traverse the space supposedly occupied by Aristotle's solid planetary spheres. It was the final blow for the ancient cosmology. But there was as yet no plausible explanatory alternative; Copernicus had been unable to supply one. The brilliant young mathematical astronomer, Johannes Kepler, who took over the mass of observational data on planetary motions that Brahe had left behind, was determined that his astronomy would not just describe the planetary motions, as Copernicus' new heliocentric model had done, it would also specify their *causes*, as Copernicus had been unable to do. And the physics would be *built* on the mathematics. But how was one to understand the elliptical motions of the planets around the sun, his first and most famous discovery? By postulating forces of attraction (he suggested) analogous to those exercised by a magnet on iron filings. In successive works, he tried to make his ideas more precise but never quite succeeded in hooking the physics to the mathematics. What he lacked, from the perspective of later mechanics, was the principle of inertia, and so he kept trying to discover what it was that urged the planets onwards on their way.[9]

His younger contemporary, Rene Descartes, had already grasped the principle of inertia, but had no use for anything like attraction. It did not *explain*, he insisted. Since only contact action could account for the movement of the planets (here he agreed with Aristotle), he proposed a fluid medium, the ether, whose eddies carried the planets along. True, this ether could not be observed. Nor could its effects on the planets be quantified in any way. So there was no way to link the physics of the ether vortices with mathematical astronomy. Once again, the same old separation. Had Descartes been asked by someone skeptical of his claims for an ether—and many were skeptical—how do you know?, his answer would have been: an ether is required by the first principles of mechanics, by the need to make movement intelligible by invoking contact action. That he was unable to link the ether with any specific predictions as to what its effects would be on the bodies it was supposed to move apparently did not perturb him. The primary criterion of a cosmological theory for him, as for Aristotle, was its explanatory coherence, its fidelity to an intuitive principle of natural intelligibility.

The cosmology of the mid-seventeenth century was dominantly Cartesian, with an ether replacing the spheres. The mathematical astronomy of Copernicus, though elegantly computational, could not even *attempt* an explanation as to why the planets moved as they did. So one still had to make a choice between computation and explanation, just as those with cosmological interests had been forced to do for near on two thousand years before. Galileo had his own way of dealing with this uncomfortable dilemma. He simply laid aside both computational and dynamic concerns, and focused on the issue of which went around what.[10] He was able to show that the ordering of the planets favored by Aristotle and Ptolemy could not be fitted to the observations of the phases of Venus supplied by his telescope. The sun, not the earth, was definitely at or near the center of the orbit of Venus. He could also show that the arguments against the earth's motion based on the physics of Aristotle were not well-founded. He could make a case for saying that a single physics applied to earth and to the moon and planets. But what *was* that physics, and how in particular was the motion of the planets to be explained? Galileo, like Descartes, could not accept Kepler's idea of an attraction between the sun and the planets. Attractions, he said flatly, explain nothing. But what was left? Galileo's spokesman in the great *Dialogue on Two Chief World Systems* remarks in response to a criticism: I do not know what it is that makes the earth and the other planets move as they do. But if my critic can tell me what it is that makes bodies on earth fall as they do, then I will tell him what it is that moves the planets.[11] So Galileo already had an inkling that a common explanation should be found for falling motion on earth and planetary motions in the sky. But he was blocked on moving further with this idea, since attraction for him was nothing more than action at a distance and thus unintelligible.

Without a dynamics, he could not prove the earth to be in motion around the sun. The best he could do against someone like Tycho Brahe who had suggested a cosmology in which the sun moves around the earth, carrying the remainder of the planetary system with it, was to urge the unlikelihood of being able to give a dynamical account of such a system, with enormous bodies rushing around a relatively tiny central body at immense speeds and maintaining perfect order in so doing.[12] Galileo almost certainly knew that a decision could not be made between the Copernican system he was defending and the Tychonic one on the basis of planetary observations only; on that score, the two systems were, in fact, mathematically equivalent. So he had to stress the advantages in *simplicity* of his own view. It would ultimately yield a simpler dynamics, he implies, a simpler description of the motions of the sunspots, and so on. But what weight could be attached to this criterion? Galileo was well aware that others would find it of dubious value. But he appealed to the intuitions of his readers: don't you *see* (he keeps implying) that this gives us a better account?

I am going to break off the story of early cosmology at this point, though it would be tempting to continue it a little further in order to recall the great debate occasioned by Newton's use of the suspect notion of attraction, not merely to describe, but (as he hoped) to *explain* planetary motion as well as the motion of fall.[13] But by now, I hope, the moral I want to draw should be clear. Cosmology of its very nature demands extrapolation, often quite daring extrapolation. Because its objects are distant and unfamiliar, it has always had to rely on indirect and sometimes precarious modes of reasoning. The divisions between Aristotelian and Ptolemaic astronomy were in the first instance about *criteria*: how should an astronomical theory be *judged*? Ought one lay weight on practical success in prediction or on explanatory appeal in familiar causal-mechanical terms? The deep divergences between

Kepler, Galileo, and Descartes in regard to the system of the world bore on how cosmological argument should proceed. How much weight ought be given to intuitive constraints on explanatory alternatives, such as those imposed by Descartes' principle of contact action? Can simplicity be trusted as a criterion of truth in a cosmological theory? There were no simple answers to questions like these then; indeed, there are no simple answers to most of them now.

The problem is that there are no accepted second-level criteria to which one could have recourse in attempting to adjudicate debate about the first-level ones. One can ordinarily decide between two rival theories in physics, say, if the criteria of evaluation are more or less agreed upon. But what if there is a debate about the propriety of relying, for example, on intuitive principles as to what counts as a "good" explanation? How is such a debate to be resolved? One may look to a past record of success (or lack of it) of principles of that sort. Or one may construct a defense, or a critique, of the recourse to intuitive principles in general. When cosmologists in the past have reached out in radically new directions, the problem has most often been to decide what criteria the new and unconventional theory should have to meet. Supporters and critics are likely to disagree even more about the criteria than about the theory itself.

## 2.   *The assessment of physical theory*

It is time now to look at the criteria that have been suggested over the years for the assessment of physical theory generally. We have already seen some of them deployed in early cosmology. But the issue of theory-criteria did not become acute until the mid-seventeenth century. Prior to that, science was expected to be *demonstrative,* issuing in definitive knowledge-claims. One began from premises that were seen to be necessarily true in their own right, and proceeded to draw conclusions deductively from them. The model, of course, was geometry, where axioms generated theorems in an orderly array. But the new sciences of the seventeenth century could not be forced into that pattern, except perhaps for mechanics. Cosmologists, as we have just seen, always had a problem with divining the nature of distant and unfamiliar objects. With the advent of the telescope, a whole new set of problems presented themselves. What were the dark patches on the lunar surface? Galileo produced a persuasive argument for calling some of them mountain-shadows. One of his arguments was: if they are caused by mountains, one would expect them to lengthen and shorten throughout the lunar day. And they do indeed do that. And no other explanation seems at all plausible. So we can conclude with some assurance that they are, in fact, shadows cast by mountains.

Take a closer look at this familiar form of argument. It is *hypothetical*: Galileo postulates the hypothesis that the lunar surface is similar to that of earth, and thus that there would be mountainous features. It is *consequential*: the warrant for the claim that there are mountains on the moon is not the self-evidence of a geometrical axiom but the verified consequences drawn from it. It is *retroductive*: it works backwards from observed effect to unobserved cause. What made this particular retroductive inference relatively simple was that mountains are familiar objects; their properties, such as the way they cast shadows in sunlight, are well-known. But in other contexts, Galileo was not so lucky. What are sunspots? What are comets? He got the first of these more or less right, and the second entirely wrong. These were not familiar kinds of things. He had to *guess* what they might be like; he had to construct a plausible model in each case and draw testable consequences from it. This was the kind of retroduction that the new scientists of the seventeenth century found problematic. What was to guide one in inference of this sort? What sorts of constraint ought be laid upon it? One was, essentially, *inventing* a cause.

Might not other invented causes account equally well for the same effects? How could one ever *exclude* that possibility where one is inferring from the observed to the unobserved?

The challenge was even more pressing when the barrier was *size*, and not distance. The new science was resolutely corpuscularian, or as we would say, atomist. (The scientists of the time were more careful than we are about the use of the term, 'atom', which means indivisible. They did not want to commit themselves to holding that the tiny constituents of which they believed bodies to be composed were strictly indivisible.) It was almost universally believed that the properties we observe bodies to have can be explained by supposing these properties to be caused by the mechanical properties: size, shape, mass, motion, of the corpuscles composing the bodies. But since these corpuscles are not only unobserved, but so far as we can tell (and unlike lunar mountains or sunspots) in principle unobservable, retroduction to them seems hazardous, not to say illegitimate. And the same sort of difficulty arose when the barrier was the more general one of non-perceptibility. We have already seen how shaky was Descartes' inference to an omnipresent ether, a substance we cannot perceive and which has to be allotted some very odd properties. And there were light-rays, and magnetic influences, and much more.

What was one to do? To survey the answers given to this urgent question by the natural philosophers of the seventeenth century would take much more time than we have.[14] But one theme recurred over and over, and is easy to separate off. If the basic form of inference in the new science was to be hypothetical, or to use our more precise term, retroductive, then the listing of the criteria proper to such hypothesis should take the place of the older and simpler logic of the deductivist tradition going back to Aristotle. Descartes, who was one of the first to realize the consequences of the shift to retroduction, still hoped that certainty might yet be reached, either by setting out all possible causal hypotheses and excluding all save one, or by erecting such an elaborate structure of tested consequences "that it would be an injustice to God to believe that the causes of the natural effects which we have thus discovered are false."[15] Elsewhere he notes "how many things concerning the magnet, fire, and the fabric of the entire world have been deduced here from so few principles" and concludes: "it could scarcely have occurred that so many things should be consistent with one another, if they [that is, the principles] were false."[16] So he was calling on what we may call the *scope* of the theory as a strong argument in its favor. Others were not nearly so sanguine. Huygens and Locke, among others, argued that the inference from observable to unobservable could never be better than probable.

One thing, however, all were agreed on. It was not enough to claim in support of a theory that it simply saved the phenomena, that all of the relevant data could be derived from it. Empirical adequacy, to use a modern label for this criterion, is much too weak, even though it may serve to exclude, or at least to weaken, some of the alternatives proposed. Kepler, writing in 1600, was already keenly aware of this, and indeed, as we have seen, the notion that saving the phenomena of *itself* could never be a sufficient testimony of truth was already an accepted principle in the earlier history of astronomy. He sought for other criteria that might help limit the field of alternatives. If medical practitioners are to be believed when they argue to the hidden causes of disease on the basis of the visible symptoms only, he says, why should not astronomers also be believed when they infer to the causes of celestial phenomena?[17] Giving a plausible *causal* account will effectively mark off one astronomer's hypothesis from another's.

There are many features of the planetary motions which had to be arbitrarily *postulated* by Ptolemy whereas they are *explained* by Copernicus.[18] For example, the superior planets are brightest and therefore at their nearest when in opposition to the sun, that is when they appear on the opposite side of the sky to the sun at sunset or sunrise. These planets also appear to move backwards in their courses at a certain point for a short time. Both of these phenomena follow necessarily from the heliocentric hypothesis: if the earth is moving around the sun, then the planets outside it will necessarily be at their nearest when on the same side of the sun as the earth is, that is, when in opposition to the sun. And they will appear to move backwards at the point where the earth is "lapping" them on the inside as both move around the sun. So there is a causal explanation for these phenomena in the system of Copernicus; there is no explanation for them in the system of Ptolemy. They are simply additional postulates of this latter system. Likewise, Ptolemy had to postulate a one-year epicycle period for each of these planets. Copernicus abolished this odd coincidence by setting the earth in motion with a one-year period around the sun. The argument here rests not merely on an appeal to simplicity or to aesthetic factors. It is a *causal* one: the Copernicus hypothesis can specify *why* such features would appear in a system where the earth is the third planet. Ptolemy cannot explain why they should appear in a system with the earth as center. He has to add them in as *ad hoc* constraints.

Kepler was convinced that a false hypothesis will, as he puts it, ultimately "betray itself" by the necessity of adding *ad hoc* features in order to keep it in conformity with new data. It may succeed initially, but

that which is false by nature betrays itself as soon as it is considered in relation to other cognate matters; unless you would be willing to allow him who argues thus to adopt infinitely many other false propositions and never, as he goes backwards and forwards, to stand his ground.[19]

Kepler was merely trying to discover the true orbits of the planets on the basis of their apparent motions and brightnesses. As the seventeenth century progressed, retroduction, inferring hypothetically from observed effect to unobserved cause, became much more complex, as the postulated causes grew less and less familiar. Robert Boyle, in his celebrated experiments on the air, tested the notion of a "sea of air" that lies all round and above us.[20] Critics argued that the apparent vacuum produced by Torricelli with his inverted column of mercury, which formed the original basis for the sea-of-air hypothesis, could not be a true vacuum, since in their view a true vacuum cannot occur. Boyle tested the two alternative views by drawing consequences from each, and concluded that the modifications that the opponents of the vacuum had been forced to introduce to explain the behavior of the mercury were not only *ad hoc* but could be more or less directly refuted. He had to introduce a new concept to designate a variable property of the air, *spring* (or as we would put it, *pressure)*, in order to account for his data And he speculated about the substructure that would have to be attributed to air—did it consist of tiny particles like coiled springs, for example?—which would explain its great compressibility. But he concluded that such hypotheses were too remote from possible test—a consideration that had never deterred Descartes and his followers—and so he left the issue aside.

But the question of how explanatory hypotheses ought to be evaluated continued to preoccupy him. In a short unpublished essay, "The requisites of a good hypothesis," he laid down six criteria for what he called a "good" hypothesis, and four more for an "excellent" one.[21] A *good* hypothesis should account for the phenomena under consideration, should be internally consistent, and should not conflict with what he calls "manifest physical truth." An *excellent* one should enable the inquirer to make new and testable predictions; it should be simple; it should not be forced or *ad hoc* in any way; and if at all possible it should be shown to be either the *only* hypothesis that explains the phenomena or, at least, the best among the available alternatives. Boyle knew perfectly well that the best he could muster in the way of explanation of chemical phenomena were vaguely plausible stories about pointed particles entering pores and the like. They might qualify, with luck, as "good" hypotheses, but hardly as "excellent" since they did not lend themselves to test or the making of new predictions. So he contented himself for the most part with the descriptive task of cataloguing chemical reactions, leaving the explanatory task to the future.

After the sophistication of Boyle's treatment of hypothesis, Newton's brusque exclusion of hypothesis from science comes as a shock. Newton believed that by separating off what he called the "mathematical" aspects of optics and mechanics, he could deduce his science directly from experiments. If light be treated as something or other propagated in straight lines, if gravitational attraction be regarded as a disposition of bodies to move in a certain way, questions about the underlying nature of light or of gravity can be left aside and unprofitable conjecture avoided.[22] One doesn't have to worry, then, about the criteria of hypothesis; hypothesis is simply not part of real science.

And deduction from the phenomena can be judged by the normal rules of logic. Later on, Newton softened his insistence on deduction, conceding that inference from experiment requires a drawing of general conclusions by induction, admitting therefore of exception.[23] But retroduction, causal inference to what he called underlying nature, remained suspect. Not that he avoided it entirely: the *Opticks* is full of imaginative suggestions about ethers, active principles, and the like. But these are labeled "queries," and seem to be regarded as prolegomena to a later stage of *real* science when the hypothetical element will have been removed. Elsewhere he appears to allow that after "mathematics" comes a "physics" in which unobserved causes can be legitimately pursued, a stage that he himself has not reached. He never did sort it all out. It is worth pondering the thought that such a towering figure in the history of cosmology should have shown himself to be so cavalier about the sensitive issue of theory-assessment.

With the coming of the eighteenth century, a new dimension began to open up, demanding retroduction of a new and particularly vulnerable sort. Natural philosophers had already faced the challenge of the very small and the very distant, distance, that is, in space. What if the distance to be bridged in imagination is distance in *time*? The record of the rock layers visible on the earth's surface hinted at immensely long and complex periods of development. Could these periods somehow be reconstructed? There were obvious barriers: only a tiny and contingent part of the earth's past had left *any* recoverable record, and the record itself could not be rerun to enable one to test hypotheses. The most serious challenge, perhaps, came from the traditional Christian belief that the earth was only a few thousand years old, and that the rock layers (with the exception of some possible vestiges of the Flood mentioned in *Genesis*) had been created by God as they now are. How is one to weigh the relative merits of such an account against those of a naturalistic explanation based on the assumption that the evidences frozen in the rock are indeed a reliable witness of an enormously long past? Each explanation has to be evaluated in its own terms, presumably, but then how can a comparative estimate be made, since some will give preliminary credence to the literal word of the Bible and some will not?

As the detail of the geological record came more clearly into view, uniformitarians and catastrophists argued, often at cross purposes, as to whether the processes responsible for forming the earth's surface were still active today. Vulcanists and Neptunists debated, often also at cross-purposes, whether the primary agency of formation had been volcanic action or sedimentary deposition. As time went on, it came to be accepted that *if* one could presuppose that the laws and theories of physics and chemistry, as these are accessible to us today, are applicable to the processes of those distant times, then one could tentatively reconstruct what went on then, whether or not the specific processes then operative are still going on at present. Even if terrestrial conditions were altogether different then, as long as a single web of laws and theories links those conditions and their outcomes with the processes of today, retroduction can bring into focus a past as distant as the record stretches.

We take this principle for granted today—and it is, of course, foundational for recent cosmology—but it is well to remember that it was not always thought quite so obvious as it now seems. The change is due to the cumulative success of the historical sciences, of geology, of paleontology, and of evolutionary biology. Success is not measured here as it might be in physics or chemistry. It is a matter of *coherence* rather than of novel prediction. The coherence lies not just in the particular historical reconstruction of a long-past geological or biological episode but in the ways in which one reconstruction supports another, and the scope of the concepts and explanatory concepts on which the reconstruction is based gradually widens. In particular, when reconstructions of quite different sorts of evidence drawn from geology and evolutionary biology, say, begin to "jump together," as it were, begin to blend fairly harmoniously into a single story, then our conviction grows that the story is not *just* coherent but is also close to the truth.

A Cambridge philosopher-scientist of the mid-nineteenth century, William Whewell, had much to say about the sort of confidence that this kind of "jumping together," or "consilience" as he named it, can give. There are, he says:

> two circumstances which tend to prove, in a manner which
> we may term irresistible, the truth of the theories which
> they characterize: the *Consilience of Inductions* from dif-
> ferent and separate classes of facts, and the progressive
> *Simplification of the Theory* as it is extended to new
> cases.[24]

Whewell combed the history of science, in a way none had done before him, for indications of how theoretical disagreement had in the past been resolved, and what the most fruitful criteria had proved to be. In a famous passage he wrote:

> No example can be pointed out, in the whole history of science, so far as I am aware, in which this Consilience of Inductions has given testimony in favour of an hypothesis afterwards discovered to be false. If we take one class of facts only, knowing the law which they follow, we may construct an hypothesis, or perhaps several, which may represent them: and as new circumstances are discovered, we may often adjust the hypothesis so as to correspond to these also. But when the hypothesis, of itself and without adjustment for the purpose, gives us the rule and reason of a class of facts not contemplated in its construction, we have a criterion of its reality which has never yet been produced in favour of falsehood.[25]

This assessment would, I suppose, be seen today as far too sanguine, as cutting truth and falsehood apart a little too cleanly. But a weaker version would still, to my mind, have merit. Whewell is arguing that one must follow the career of a theory in order to assess its credibility as a claim to knowledge. One ought not merely ask: is the theory empirically adequate, here and now? One has to ask: how well has it guided research in the past? Has it given rise to novel predictions, later verified? Has it unified domains hitherto seen as disparate? Has it suggested means of overcoming anomalies that have arisen? Has it avoided *ad hoc* modifications in the face of apparent anomaly? Has it led to significant theoretical simplification? To evaluate a theory, then, is in part to determine how well it has served as a research program up to this point. To the extent that the theory is true, that the causes it postulates exist as the theory describes them, one would *expect* it to guide research effectively. To the extent that it is not true (and all sorts of degrees occur here), one would expect it to fail, at least occasionally, as the basis for a fruitful research program. Modifying Kepler slightly, the falsity in a theory will tend to betray itself eventually.

Those of you who are familiar with recent philosophy of science will have caught echoes of a controversy which has been raging for several decades now. The controversy concerns what is called scientific realism, the claim that the continued success of an explanatory theory warrants some degree of belief in the existence of the entities postulated by the theory. Put like that, it seems a harmless claim, and certainly one to which most working scientists would subscribe. Why should anyone ever question it? The challenges here, in fact, come from many sources.[26] Questions arose in physics, for example, about the reality of the ether; its existence seemed to be demanded by a prior principle that the transmission of electromagnetic and gravitational action required a medium of some sort, but there appeared to be no *direct* way to test for its presence. Atoms for a long while were suspect, partly because there seemed to be an alternative formalism, that of energetics, which accomplished most of what atomic theory did, but even more because empiricist scientists and philosophers could not take seriously an existence claim for entities that in principle could never be observed. The source of malaise at present lies, of course, in quantum mechanics. It is no accident that most of the leading anti-realists today are specialists in the philosophy of quantum mechanics. Defenders of realism respond that there are, undoubtedly, difficulties in deciding what ontological status to attribute to electrons, quarks, and the like. But they insist that this in no way weakens the claim of realism in less exotic contexts like chemistry *or* molecular biology.[27] And even in the quantum realm, realists argue, the problem is not *whether* elementary particles exist but just how to *interpret* them in categories we can grasp.

By far the main challenge to realism comes, however, from those who follow the lead of Thomas Kuhn's influential book of thirty years ago, *The Structure of Scientific Revolutions*. Kuhn emphasized the importance of revolutions, of what amount to discontinuous theory-changes in the history of science, and inferred from this that the succession of theories in any given domain ought not be taken to converge on the real structure of that domain. Later theories may be better puzzle solvers than earlier ones, but this alone does not permit us to conclude that they also come closer to telling us what some part of the world actually consists of. Though the same *word*, "electron," for example, may carry through theory-change after theory-change, the changes in meaning it undergoes are so great that we must not allow ourselves to suppose that there is something called an "electron" we are coming to know better and better.[28]

The issues here have proved to be very complicated, and I am certainly not going to be able to rehearse them all here. But one point worth emphasizing is that the anti-realists' refusal to allow explanatory success to justify belief in the existence of specific underlying causes would strongly affect cosmology. The critics of realism tend to focus on the history of mechanics, on the shift from Newton to Einstein in particular. How do you know, they ask that another conceptual shift equally far-reaching may not lie ahead, in which Einstein's theories are superseded as thoroughly and definitively as Newton's were? But the cataclysmic events of the distant past postulated by cosmologists are just as theory-dependent as electrons or forces are. If we believe that these events are likely to have occurred, it is because the cosmological theory in which they appear as causes of a train of events stretching to the present is recognized as a *good* theory, as what Boyle would have called an "excellent" hypothesis. The critics *cannot* both hold that the succession of theories in physics and chemistry improve on one another only as problem-solvers, *and* also hold that the sciences of the past, cosmology, geology, paleontology, tell us something worthy of belief about what has happened down through the ages. If we are not licensed to believe, on the basis of explanatory success, that physics gives us an insight into what goes on beneath the level of direct observation, we are for the same reason barred from belief that cosmology allows us to reconstruct events no longer directly accessible to us.

The realist response to the anti-realist challenge has taken many forms.[29] There is time to sketch only two of these. Retroductive argument has in the past often led to the postulation of causes whose existence was later more or less directly shown. Galileo's lunar mountains immediately come to mind, but there are plenty of examples in recent microphysics where sophisticated new experimental methods of probing, say, solid surfaces, have made it plausible to claim that atomic-level entities are being in a quite definite sense "observed." The same form of argument that led in the first place to the postulation of these now well-attested entities is used in other more problematic contexts, like cosmology. There is no reason to suppose that the criteria of explanatory success that have proved themselves in one case fail in the other, even though they may have to be employed more cautiously in some contexts than in others.

A second argument can be based on the progressive history of such sciences as organic chemistry, cell biology, and structural geology. In each of these cases, a model-structure initially provisional has been filled in with more and more detail. There has been a steady development, where each stage builds on those before.[30] There have not been the sharp ruptures that Kahn and Bachelard have singled out, where the later explanatory scheme seems to jettison some of the central features of earlier ones. Atoms have taken on more and more flesh, as it were, since they made their first appearance in Dalton's chemistry and Avogadro's gas theory early in the last century. Only *very* rarely has a structural element that has been explanatorily fruitful over a considerable period been rejected later. What ordinarily happens is that an element first appears as a near-blank which is gradually filled in as time goes on, as theory is revised and reworked in the light of new and often anomalous data. The notion of electron orbit, so intuitively simple in the original Bohr model of the hydrogen atom, has been gradually transformed in the light of later developments in quantum theory. But it is not difficult to trace a continuity, a sense of progressive clarification, in which part of the original insight is retained, and the insight itself greatly deepened.

This is, I should emphasize, a more optimistic assessment than many of my colleagues in philosophy of science would be willing to allow. And it would need far more detailed historical argument in its support than I have given it. Furthermore, it would now be challenged from a new quarter also, as the newly-developing field of sociology of science produces arguments for seeing science as a social construction, an arena where competing political and social interests play a crucial part in theory-acceptance.[31] Nevertheless, I think that the realist case can be sustained. The great contribution of anti-realism has been to underline how carefully the realist thesis has to be circumscribed, how tentative must scientific claims that are based on explanatory success be regarded, and above all how sensitive one must be to the *differences* between explanatory theories. There is no *global* realist thesis; each theory has to be separately assessed, not by applying some sort of philosophical yardstick but by attending to the criteria that have guided scientists and intrigued philosophers, from Kepler and Boyle down to Whewell and Kuhn.

### 3. *Cosmology lays claim to the status of a science*

Only a century ago it would have seemed quite impossible to extend our knowledge backwards in time in cosmology as scientists had already begun to do in geology and evolutionary biology. In geology there was a detailed, if often confusing, record of geological change locked up in the rock strata. In evolutionary biology, there was a novel theory which offered a means of interpreting contemporary organisms and their traces in the ancient rocks. But in physics, Newton reigned. And according to the *Principia,* space and time are without bounds of any sort. Space is infinite, and time is without beginning or end. In such a universe, solar systems can come to be, but there can be no real evolution on a cosmic scale. There is no way, in fact, to formulate a real cosmology, since nothing can be said about the extent or duration of matter. One might debate whether or not it also is infinite in quantity, but beginnings and endings are radically out of reach. In Newton's own view, these voids could only be filled by theology. If the stars of the night sky had a beginning, we could know of it only through God's direct revelation.

Two things changed all this in a matter of decades. The first was Einstein's formulation of the general theory of relativity, and his application of it in 1917 to the construction of a universe-model. The melding of space and time into one and the substitution of non-Euclidean for Euclidean geometry enabled him to create a matter-filled cosmic model that could be at once finite and unbounded, and thus able to escape the paradoxes that philosophers had in the past urged against the notion of a finite universe. Wilhelm de Sitter immediately went on to show that a model empty of matter could also be constructed that would satisfy Einstein's equations, but had a very odd property that test-particles placed in it would move away from one another with ever-increasing velocity. Finally, in 1924, Alexander Friedman constructed a whole family of relativistic universe-models in which matter-density varied with time; some expanded, some contracted, some remained steady. The theoretical groundwork had been laid for the big bang.

Then came Hubble's discovery of the galactic redshifts in 1929. The simplest (though, as Hubble himself emphasized, not the only conceivable) interpretation was that the galaxies are all moving away from us at great speeds, speeds that seemed to be roughly proportional to their distances from us. Lemaitre saw how theory and observation were converging, and proposed that the galactic motions could best be understood as an expansion of space itself, consonant with Einstein's equations. Here was the constraint that was needed to pick out one among the Friedman models. Since the galactic redshifts could not only be explained after the fact in terms of a Friedman model, but could in a sense have been anticipated once the Einstein equations are applied to the physical cosmos as a whole, Hubble's discovery might be seen as offering support to the enormous inductive leap involved in extending the general theory of relativity from the solar system to the universe as a whole.

But an immediate difficulty presented itself. From the rate of expansion, one could calculate an upper limit on the length of time elapsed since the expansion began, loosely called the age of the universe, assuming only that the galaxies had not accelerated over time. Using Hubble's value for the expansion constant, the age came out as roughly two billion years. This was much shorter than the age of the solar system as that had already been estimated, using a variety of geological indicators, shorter too than the ages of the oldest stars. So the expanding universe model seemed to fail its first major test. One consolation was that it showed that at least the model was a *testable* one, that cosmology was now at the stage that, unlike Cartesian physics, it did not have to rely on a broad explanatory coherence only.

This setback was one, though not the principal, motivation for the construction by Bondi, Gold, and Hoyle of an alternative steady state model in 1948. Earlier cosmologists had proposed what they called a "cosmological principle," that the universe is relatively uniform over large distances, that the galactic distribution is, in effect, homogeneous. It was never quite clear whether this was an idealization adopted in the absence of more precise information, or whether it was being proposed as *normative*, as being a restriction on acceptable cosmological models because of its being more likely in its own right to be true, as the term "principle" suggests. The proponents of the steady state model relied on an even larger scale claim. The "perfect cosmological principle," as they called it, asserted a uniformity in time as well as in space, that is, the universe always has and always will look much the same as it does now. No privilege for the present! They appeared to attribute a normative status to the idea that our location in space and time is no way special. Others have called this the "Copernican principle, a usage that would have utterly astonished Copernicus. A better label might be the "Principle of Average Location."[32] A further reason for proposing a universe without a beginning was a conviction, one shared by many physicists of the day, that the notion of an absolute cosmic beginning was intellectually abhorrent. To account for the apparently expansive galactic motions, Bondi, Gold, and Hoyle postulated the continuous creation of hydrogen throughout space at a rate so low as to be undetectable, but sufficient to allow the galaxies, in essence, to be replaced in our field of view by other newly-forming galaxies, as the older ones move out of view.

They regarded the multitude of little bangs as preferable to a single big bang; the steady creation of hydrogen could be thought of as lawlike and thus in some broad sense intelligible. The fact that it violates the conservation of energy seemed to the defenders of steady state as simply indicating that the conservation principle ought not be extrapolated that far. If one had to choose between the principle of conservation of energy and the perfect cosmological principle, their preference was clear. In retrospect, it would seem that the proponents of the steady state theory were guided more immediately by *a priori* principles of natural intelligibility than their rivals were. On the other side, Arthur Eddington, the first to popularize the expanding universe model, never found himself able to accept the notion of a beginning in time. The Biblical concept of creation had always, of course, been taken to involve such a beginning. But there can be no question about the uneasiness among the original proponents of the big bang model about the singularity in time that their theory required. The impulse to this theory came in the first instance from the Einstein equations and the Hubble observations, even though many Christians later came to see in the new cosmology a confirmation in some sense of their own faith.[33]

The 1950's was a time when the defenders of the rival models battled tenaciously but indecisively to establish the superior merits of their own case. By this time the original objection to the big bang model had been removed, when a recalculation of the Hubble constant in 1952 showed a safe figure of ten to twenty billion years for the age of the universe, well over the estimated ages of the various cosmic constituents. Nevertheless, the defenders of the perfect cosmological principle were not impressed. Their assurance would not be challenged, they implied, unless a direct conflict with observation could be established. A mere extrapolation from observation, like the principle of conservation of energy, would not carry enough weight.

In his book, *The Unity of the Universe,* which appeared in 1961, the cosmologist Denis Sciama assessed the relative merits of the two models at that time. In his view, the steady state model was clearly superior, for one reason in particular. One aim of science, he says "should be to show that no feature of the universe is accidental."[34] Only if the universe as a whole is unchanging in time, the continual creation of matter compensating for the expansion, he goes on, will its main properties be "fundamental" (as opposed to accidental). If there was a beginning of the universe, the initial conditions would have to have been contingent. We must:

> find some way of eliminating the need for an initial condition to be specified. Only then will the universe be subject to the rule of theory. . . . This provides us with a criterion so compelling that the theory of the universe which best conforms to it is almost certain to be right. [35]

The choice is decisive in favor of the steady state model, he concluded, "because it is the only one in which the properties of galaxies and the cosmical abundances of the elements can be calculated without any accidental initial conditions. This model is the one that satisfies the perfect cosmological principle." Whether the universe really *does* satisfy this principle, he concedes, "may be decided by observation in the next few years."[36] He was right about this, but the decision did not go the way he so confidently expected.

This episode illustrates, as well as does any in modern cosmology, how risky reliance on principles of natural intelligibility has proved to be in the history of physics. From Aristotle and Descartes all the way to Hoyle and Sciama, their track-record has been poor.[37] Sciama found in his indifference principle (a good cosmological theory ought to be indifferent to the choice of initial conditions) the basic reason why the perfect cosmological principle holds good. And it had so strong a hold on him that he claimed as additional support for his preferred model that it "could explain" the properties of galaxies and the cosmical abundances of elements. There is a lesson for us here also because the "could explain" of this claim was of the roughest and most qualitative sort. Helium abundance, for example, was to be explained by a constant burning of hydrogen within stars, but it was not at all clear how this was to yield the figure for cosmic helium abundance (7% by number of atoms) that he was using. His suggestion that the known abundance of helium in the Milky Way galaxy could be accounted for by the stars "burning their fuel from the time they were formed until today" turned out to be wrong.[38] When one is relying on a persuasive-sounding principle of intelligibility as the main warrant for a particular cosmological model, one is apt to be less demanding about what counts as observational confirmation.

What came next has become the stuff of legend.[39] In 1964, Penzias and Wilson found an unexpectedly strong microwave radiation at a wavelength of 7.35 cm that appeared to be coming uniformly from all parts of the sky. A fortunate interaction with Peebles and Dicke at nearby Princeton led to the recognition some months later that this radiation could well be a relic of the radiation from the big bang that Peebles and Dicke had been searching for. Nearly twenty years before, Alpher and Herman, while exploring George Gamow's theory of how the chemical elements had been synthesized in the early universe, had forecast that there should be an omnipresent microwave radiation with a temperature "signature" of around 5 K [40] But Gamow's theories of nucleosynthesis had not worked out for elements heavier than helium and the big bang model from which they were derived was under increasing attack at the time, so that the forecast had been forgotten. Peebles, who was unaware of the earlier forecast, had himself predicted a "relic" radiation with a temperature signature of around 10° K.

The success of these forecasts proved to be a critical turning-point in the history of modern cosmology, for several reasons. The fact that a very low-temperature uniform microwave radiation had been seen in advance to be a likely consequence of the big bang model now counted enormously in favor of that model. A novel prediction had been more or less verified; an effect until that time unnoticed had been predicted well in advance. This was, in logical terms, just the sort of confirmation that scientists from the time of Kepler and Boyle onwards have seen as most persuasive. Some philosophers have argued that the fact that a theory predicts a *novel* effect, subsequently verified, gives no more support to a theory than if that effect were part of what the theory was originally constructed to explain. Either way, they say, it is a verified consequence of the theory, no more.

But most scientists (and many philosophers) would disagree and would argue that the fact of the successful prediction's being a *novel* one gives additional credence to the theory. It lessens the worry that the theory may be *ad hoc,* a testimony to scientific ingenuity rather than to the truth of the matter. If the theory is true, or at least approximately true, one would *expect* it to give rise to successful novel predictions; one would *expect* it to be fruitful over the course of time. Or to put this in a different idiom, one would expect it to survive observational test. Whereas if the novel results had been built into the theory from the beginning, it could not be said to have been *tested* to the same degree.

The discovery of the 3-K radiation not only was taken to be a validation of the big bang model, but virtually eliminated its main rival. The homogeneity of the radiation indicated that it did not come from our own galaxy, that it must have come from an earlier more homogeneous universe, and thus that cosmic evolution must have occurred. The steady state model, which excluded the possibility of such an evolution, was thus undermined. It was difficult (some said impossible, though one should never underestimate the ingenuity of theorists) to account for such a radiation if the universe has always been more or less as it now is. In 1971, Sciama wrote that the steady state model had been more or less eliminated by the new evidence.[41] One point worth noting here is that there were two, and only two, families of solutions to the problem of cosmic origins, one of them holding that the universe has always been as it now is and the other, that it has become what it now is, beginning from a finite time in the past. The virtual elimination of one family of solutions immensely strengthens the case for the other, much more than would be the case were there to be a large number of possible solutions, as is more often the case. This was, as we have seen, the method above all others that was recommended by the theorists of the new science of the seventeenth century. But it is rarely available.

4. *Cosmic evolution from an initial highly condensed state: How do we know?*

Finally, let us review very briefly the lines of evidence supporting the "standard" big bang model, lines of evidence to which the norms of theory assessment I have already sketched can be applied. [42]

First, the theory obviously conforms to the equations of general relativity and the evidence of galactic expansion on which it was originally based. Second, the application of quantum theory to the early moments of such a cosmic expansion leads to a prediction of an expanding radiation, filling space and gradually cooling. Such a radiation would have the distinctive "signature," or energy distribution, of a "black" body, one whose matter is in equilibrium with the surrounding radiation. Penzias and Wilson supplied only a single point on the blackbody curve; since then a great deal of effort has gone to checking the curve at shorter wavelengths by observations which have to be made above the shielding of the earth's atmosphere. And the frequency distribution turns out to be more or less exactly thermal or blackbody. It is hard to see how a thermal radiation so exactly homogeneous could have originated except in an early homogeneous very hot universe. Steady state models would simply have to postulate such radiation as an "add-on," as Ptolemy postulated the one-year periods of the epicycles.

A somewhat weaker argument relies on the various methods by which the age of the universe can be calculated. The phrase "age of the universe" is a bit misleading in this context. All it means is the time elapsed since the cosmic expansion began. [43] Since Hubble's day, thousands of galaxies have been charted and their redshifts measured. The results are best explained in terms of an extraordinarily uniform expansion, one that began between ten and twenty billion years ago; the breadth in the estimate is due to our ignorance of whether or not the expansion has been gradually slowing. If the universe is open, the age would lie at the twenty billion end of the scale; if it is closed, the preferred figure would be closer to thirteen billion. Independent methods of dating suggest an age of about four and a half billion years for our solar system, an age of perhaps ten billion years for the oldest radioactive elements, an age of close to fifteen billion (it seems) for the oldest globular clusters. So there is a broad consistency, at least, in the various age estimates.

Besides the cosmic background radiation, are there any other features of our universe that might plausibly be construed as relics of an original "fireball"? Are there any other causal lines that lead back from the present to an extraordinarily condensed state of matter in the remote past? As early as the 1930's, Victor Goldschmidt and others attempted to calculate the cosmic abundance of the different elements, based on the assumption (for which there was already some evidence) that the relative abundance was fairly constant across our own galaxy, at least.[44] It was not a long step from this to the idea, first pursued by Gamow and his co-workers immediately after World War II, that a *common* abundance could best be explained by postulating an original hot dense state, a primal event in which energies would be sufficiently high to synthesize the nuclei of the different elements one after another, working progressively upwards from the simplest, hydrogen. Initial calculations for the lightest elements were encouraging, but as we have already noted, plausible transformation paths to the heavier elements appeared to be lacking.

In the decades since then, an alternative scenario for the genesis of the heavy elements has been worked out, involving processes that are believed to occur in supernova explosions. The relative abundances of these elements varies a great deal; it seems likely, therefore, that contingent processes involved in the formation of individual stars are likely to have played the major role in their formation. For the lightest elements (deuterium, helium, and lithium), however, the evidence for a uniform cosmic abundance has continued to mount, indicating a common origin in an earlier highly condensed state of the universe where the high energy-densities required for nucleosynthesis could have been found. The argument regarding helium abundance is especially impressive; it offers epistemic support to the big bang model comparable to that given by the cosmic background radiation itself. Various types of observations of various types of objects (stars, nebulae, cosmic rays, other galaxies) give a figure of about one atom of helium for every eleven of hydrogen (or a mass fraction of about 25% for helium). Theoretical calculations of the production of helium out of hydrogen in the first minutes of the big bang, using the standard parameters of the model, yield very similar figures. (Peebles, for example, gives a mass fraction of 25% ± 2.5.)[45] The computation is relatively robust against small changes in the model parameters.

The measure of agreement between the "observed" helium abundance and the abundance predicted by the big bang model is striking. Helium would also have been produced by hydrogen fusion in individual stars, but this process could (it is argued) account for no more than 10% of the observed abundance.[46] It must be stressed that a great many assumptions are involved in the measurement of helium abundances, and occasional variations from the expected abundance norm have been registered. Nevertheless, the fact that measurement after measurement of helium abundance comes up with a figure which is close to the one that the standard big bang model yields, without need for any special adjustment of the parameters, converts into strong support of the model. The support is all the stronger in that persistent attempts to find any even halfway plausible alternative scenario to an initial fireball as a means of explaining a uniform transgalactic abundance of helium have failed. Comparable arguments can be given on the basis of observed deuterium, tritium, and lithium abundances, but here the abundances themselves are so much smaller that the uncertainties in the calculations take on greater significance.[47]

Some other more indirect lines of argument ought also be noted. Evidence has mounted that the universe in the distant past was very different to the universe of today, and hence that cosmic evolution has occurred. According to the big bang model, galaxies at high redshift (hence further away, hence younger) ought to show an earlier stage in stellar evolution; there is considerable evidence for this, though there are also anomalies.[48]

More significant is that very distant galactic clusters include blue galaxies of a type that seem to occur *only* at great distances (i.e. only in the distant past, in an earlier stage of cosmic evolution). And at even higher redshifts (when the universe was only about a quarter of its present age according to the big bang scenario), quasars become the dominant type of object, being at least a thousand times more numerous than they are in more recent times. An independent argument for the crucial interpretation of quasar redshift as an indicator of distance (as well as, indirectly, for the big bang model itself) has been given by some beautiful observational evidence of gravitational "lensing" where the light of a more distant quasar is bent, in the manner Einstein predicted, as it passes around a nearer galaxy.[49] Finally, new data on radio galaxies suggests that they emitted much more radiation in the distant past, since there are many more to be observed at great distances than would be the case if they always radiated as they do now. The evidence for cosmic evolution is thus by now extremely strong, effectively excluding steady state types of theory, and supporting the view that the observable universe was generated in a high-density fireball billions of years back.

The standard big bang model appears to satisfy the criteria of theory-appraisal discussed earlier remarkably well. Three presumptive causal traces of a big bang (the galactic redshift, the cosmic background radiation and the uniform helium abundance) together carry strong conviction, particularly because every effort to find even moderately plausible alternative explanations for them has failed, and because the quantitative predictions have matched theory and observation quite neatly, without need for *ad hoc* adjustment. The big bang model itself has proved to be fertile in all sorts of ways as a research program; its scope has gradually widened, and early anomalies have been overcome. Nevertheless, challenges remain; since the epistemic strength of the model depends not only on the positive evidence in its support but on the quality of the challenges facing it, a brief word about these is in order. They are mainly of two kinds. The kind that has recently been in the news is the inability of the standard model to account for the origin of galaxies and particularly for the large-scale galactic structures that have recently been extracted, with much labor, from the evidence. So the model may have to be modified, which is what normally happens: what has been so striking about the standard model so far is that it has *not* had to be substantially modified up to this point. It is not as though there is a rival theory which *can* account for galactic origins. What has made the problem so intractable is the extraordinary homogeneity of the background radiation. The sort of inhomogeneities that would be needed in the initial stages of the cosmic expansion to explain the genesis of great sheets of galaxies surrounding vast voids[50] *ought* have affected the background radiation also. It is the combination of inhomogeneity in the present state of the one and homogeneity in the other that makes a coherent account of origins so hard to construct.

The other challenge has been not so much to the big bang theory itself but to the velocity interpretation of the galactic redshifts on which the model depends. Arp and his co-workers have claimed to find a significant number of cases where a supposedly nearby galaxy and a supposedly distant quasar seem to form part of one system.[51] Other cosmologists have challenged the statistics underlying their argument. Another hypothesis that has had intermittent support is that light may redden as it ages. There is no independent evidence for this, nor any theory which would lead one to expect it. Though it cannot at this point be excluded, the velocity interpretation of galactic redshift has by now strong indirect support from the successes of the big bang theory.

Attempts to formulate alternatives to the big bang theory have encountered a major obstacle: the cosmic background radiation. Arp, for example, has suggested that this radiation might come from intergalactic dust which absorbs and reradiates energy from the galaxies. But efforts to show how this radiation could have the thermal blackbody spectrum the 3-K radiation is known to have, have not been successful.[52] Alfven's plasma model of cosmic origins, publicized by Eric Lerner in his provocatively-titled recent book *The Big Bang Never Happened*,[53] encounters a similar difficulty in accounting for the properties of the background radiation. In addition, its reliance on a symmetry between matter and anti-matter in the early universe leads to anomalous consequences, such as a high flux of gamma rays which has not been observed. Obviously, these and other alternatives to the big bang theory have to be carefully investigated and tested, but for the moment at least, the presumption must be held to be heavily in favor of the big bang.

"For the present, at least. . . ." How durable *is* the theory? Might it be displaced by an unpredictable Kuhnian revolution at some point in the future? One can never exclude such a possibility, of course; one can only estimate, as best one can, the likelihood that the "core" of the standard big bang theory might have to be entirely rejected. The theory itself will undoubtedly continue to develop, as atomic theory continues to develop. But just as it can be argued rather persuasively that atoms are *most* unlikely to disappear at a later more sophisticated stage of theory, so the epistemic analysis above gives strong warrant for saying that the idea of a cosmic expansion from a highly condensed state in which the light elements were synthesized and a diffuse radiation given off is very likely to last.

### 5. *Postscript: The inflationary hypothesis*

Nothing has been said above about the first fraction of a second of the expansion, when the energy-destiny was so high that the four fundamental forces had not yet fully separated off from one another. It must be emphasized that this lies outside the scope of the standard big bang model, and poses questions that the model cannot answer without supplementation. Our conclusions above about the epistemic support the standard model enjoys do not extend to hypotheses about what happened in the first millisecond of the expansion. The galactic expansion, the emission of the cosmic background radiation, and the formation of helium, the processes that have furnished the primary evidence for the standard model, originated later, according to the standard model. The energy-densities during that first instant lay well above what our particle accelerators can produce and at the very boundaries of our present elementary-particle theories. Reconstructing the earliest moment of the expansion carries us, therefore, well beyond the range of our present navigation charts. The standard model, by contrast, can be counted as a relatively straightforward realization of Einstein's equations.

Elsewhere in this volume, the essays by Philip Morrison and William Fowler sketch a recent and dramatic new account of that first millisecond, of what Morrison in his retrochronology calls "Act Two." (He describes all of cosmic time back to that millisecond as "Act One.") Alan Guth's inflationary hypothesis (1980) is an addendum to standard big bang theory, a filling in of what might have happened in the first $10\text{-}3°$ seconds, with a view to relieving the theory itself of several sources of strain. To understand the genesis of Guth's hypothesis, let us return for a moment to the indifference principle that we saw Sciama rely on so heavily. It runs: whatever theory we propose for the early universe, it ought to be indifferent to (independent of) any particular initial conditions.[54] Charles Misner's "chaotic cosmology" in the 1960's took this principle as its starting point, but was unsuccessful in finding a way to smooth out initial anisotropy.

The principle itself came under severe challenge when in 1973 Collins and Hawking calculated that the only way to get a long-lived universe like ours, having an energy density close to the critical value that constitutes the borderline between an "open" and a "closed" universe, was to have an initial energy-density at almost precisely the critical value.[55] Not only was the choice of initial conditions not indifferent, as the principle prescribed, a degree of precision in the initial setting was needed that seemed to defy the odds. At this point, Collins and Hawking introduced the notion that Brandon Carter would later call the "anthropic principle," suggesting that finding the universe to be flat can be understood as "a consequence of our own existence: if it had not been flat, we wouldn't be here."[56] That alone, of course, would not explain anything, but the authors added the further suggestion that if one adopted a many-universe model besides, so that other values of the energy density are realized in other universes, we *of course* would be in the one that is flat. A quite different mode of explanation suggested itself to those who saw the universe as depending on a Creator for its existence; since according to the Biblical tradition, human beings play a special role in God's design for the world, it might be supposed that the fine- tuning of the initial conditions could be explained straightforwardly (though not scientifically) as a choice on the Creator's part. In the late seventies, the debates over the "anthropic principle" in all its varied forms were vigorous but indecisive.[57]

Guth's inflationary hypothesis can now be seen as in part an effort to restore the indifference principle to its former primacy.[58] By postulating an enormous expansion prior to 10-30 seconds, one large enough at 10-50 to inflate an atom to the size of the observable universe, it is possible to smooth out any initial isotropy entirely. No constraints need to be laid on the initial energy density, and the indifference principle is once again safe. In addition to eliminating the flatness problem which had helped to give rise to the earlier claims for fine-tuning, Guth's model also removed an anomaly of a rather different logical sort in the standard model. The "horizon problem," as it was called, referred to the fact that in the first microsecond, the component regions of the nascent universe would be separating so fast that causal co-ordination between the regions could not be maintained; yet there is evidence (notably the background radiation) that it *was* maintained.

According to the inflationary hypothesis, the region from which our observable universe evolved is so much smaller than the corresponding region in the standard model that causal connection *can* be maintained. But there are still horizons: the observable universe is now only a micro-region in a vastly larger maxi-universe. And the multitude of other microregions have been out of causal contact with ours since the first instant of the cosmic expansion and will forever remain so. Guth has simply pushed the troublesome horizons further back in order to secure causal connection at the origins of our own observable universe, where we know it to have been necessary. Obviously, no direct test of the existence of these other "mini-universes" is possible. The only warrant for such an existence-claim is the explanatory coherence of the larger theory of which it forms an integral part.

Since it was first suggested by Guth, the inflationary hypothesis has been amplified and revised by Steinhardt, Linde, and others.[59] Malaney and Fowler have shown a possible way round an important objection to the inflationary model regarding the necessity in this model of postulating a large amount of "exotic" non-baryonic matter (for whose existence there is absolutely no evidence at the moment).[60] The details of these arguments lie beyond the scope of this essay. What makes the inflationary hypothesis especially relevant to our theme is the epistemic contrast between it and the standard model, as this latter applies to cosmic history subsequent to the first second or so of the expansion.

The inflationary hypothesis must be adjudged as yet rather speculative. There are no clearly identified causal links to the supposed inflationary era enabling specific predictions to be made. "Grand unified" theories which would govern the energy regime prior to the separating off of the strong nuclear force and the electroweak force are still quite tentative. These theories do allow for phase transitions of the sort that inflation requires, but general consistency falls far short of testable prediction. Malaney and Fowler note that the theorist who is engaged in constructing a model of this first expansionary moment has the "luxury of a large parameter space," made even larger by the introduction of the superstring formalism.[61] It cannot be said that inflation is required by the Einstein equations, though it can be made consistent with them if one adds in, as Einstein originally did, a cosmological constant representing a cosmic repulsion. But Einstein did this in order to *inhibit* cosmic expansion, not to facilitate it. And the term is an arbitrary one, *ad hoc* in the sense in which some features of Ptolemy's formalism were *ad hoc*.

The inflationary hypothesis rests on a single type of argument. The standard model encountered problems in dealing with the first moment of the cosmic expansion. Inserting an inflationary phase within that first moment removes (or at least lessens) those problems. Causal horizons in the early universe that would be at odds with the observed homogeneity of the background radiation are simply pushed further out so that we can no longer observe their effects. Monopoles and domain walls are not observed by us. So push *them* back also: inflate the universe to such a vast degree that the region accessible to our instruments is left as only an almost infinitesimally small part of the universe as a whole. The chances of encountering monopoles or other awkward features are correspondingly decreased: they are still there, but vastly more spread out. The epistemic price is high, though. Most of the universe (nearly all of it, in fact) has been out of causal contact with our observable region since the first instant of cosmic expansion, and will forever remain sundered from us. Causal links of the sort that confirmed the original big bang theory thus appear to be in principle excluded.

A consideration that carried much weight with Guth was the elimination of the "flatness" problem. The original big bang theory seemed, as we have seen, to require a "fine-tuning," a very precise constraint on the initial cosmic parameters, in order that the universe be of the long-lived sort that would permit heavy elements to be formed, planetary systems to develop, and biological evolution to occur. This ran counter to the indifference principle that has carried so much weight in the recent cosmological thinking. The inflationary hypothesis makes fine-tuning of the original energy-density unnecessary, since the inflation will force the energy-density towards the initial value no matter what its original value was.[62] This is enormously attractive to cosmologists since it eliminates the apparent contingency of the cosmic beginnings. But what is the epistemic force of the indifference principle, and from where does it derive? As an operating assumption, it facilitates scientific inquiry, true. But is there any antecedent reason why the universe *has* to be that way? Guth and Steinhardt remark: "Most theorists (including both of us) regard such fine-tuning, the degree of fine-tuning that still seems to be required in the inflationary model itself as implausible."[63] But on what grounds? Calling on "convictions," "hunches," and the like may not serve as any more trustworthy a guide than it did for the proponents of the principle of contact action in earlier mechanics.

In their review of the current evidence in favor of the big bang model, Peebles and his co-authors are careful to separate off inflation as an independent hypothesis. Their verdict:

> What do we do about inflation? The division of opinion among the authors of this review may be a reasonably close reflection of opinion in the community. Half of us consider inflation attractive, and would deeply regret its loss if it were somehow shown to be wrong. The other half wonder why we are care so deeply about a theory that makes so few definite testable predictions. Perhaps quantum gravity (or string theory?) will one day find an alternative to inflation for setting the cosmological initial conditions. But pending further growth of understanding of physics under extreme conditions, we live with inflation, as one of the open puzzles.[64]

To sum up, then, our lengthy survey of the early history of cosmology and of the criteria that gradually came to govern theory-choice in natural science would lead us to assert with a fairly high degree of confidence that our universe is the product of a vast cosmic expansion that began somewhere between ten and twenty billion years ago, as this is described in standard big bang theory. But the sequence of events in the first fraction of a second is still unclear. A "roll-over" phase transition leading to a mind-numbing inflation by a factor of 10-50 or more is an hypothesis worth serious investigation; it has already had some success in providing favorable initial conditions for the "standard" big bang to get under way. It is too early to tell how successful a research program it will engender, as grand unified theories develop. And there is some reason to question whether the causal links to that first instant will ever prove as distinctive as those that, at energies closer to those we can reproduce, have afforded strong testimony to the "big bang" version of later cosmic history.

## NOTES

[1]William E. Schmidt reported on this story in the *New York Times* of September 1, 1991, p. A9. The historian, John Bossy, of York University is the author of *Giordano Bruno and the Embassy Affair*, Yale University Press, to appear Fall 1991.

[2]See Mc Mullin, "The goals of natural science", *Proceedings American Philosophical* Association, 58, 1984, 37-64.

[3]See Otto Neugebauer, *The Exact Sciences in Antiquity*, Providence: Brown University Press, 1957, p. 115. Two different formalisms *were* in fact used by the Babylonian astronomers to compute the dates of lunar eclipses. The older one, System A as it is sometimes called, employs the equivalent of an arithmetical step-function. The later one, System B, uses a more complicated zig-zag function. System A is somewhat easier to use; System B is somewhat more accurate. It is interesting that both systems remained in use throughout the heyday of Babylonian astronomy (250 B.C. to 50 B.C.). An even greater diversity of numerical methods was employed for planetary calculations, offering several different methods to compute the same phenomenon.

[4]For a nuanced account of Greek cosmology, see the works of G.E.R. Lloyd, notably *Magic! Reason and Experience,* Cambridge: Cambridge

University Press, 1979, chaps. 1 and 3; *The Revolutions of Wisdom*, Berkeley: University of California Press, 1987, chap. 5 and 6; *Early Greek Science: Thales to Aristotle*, London: Chatto and Windus, 1970.

[5]Quasi-mechanical because there are actually two sorts of agency at work in Aristotle's system: contact action between the spheres, as each sphere carries the poles of the sphere within it (which do not coincide as a rule with its own poles), and teleologically guided action, as a soul or living agent is required to give each sphere its own characteristic rotation. For more detail on this very complex system, see Lloyd, *Ancient Greek Science*, pp. 86-94, and Mc Mullin, "Realism in the history of mechanics", to appear.

[6]See Lloyd, *Magic. Reason and Experience*, pp. 175.

[7]See Mc Mullin, "The explanation of distant action: Historical notes", in *Philosophical Consequences of Quantum Theory*, ed. J. Cushing and E. Mc Mullin, Notre Dame: University of Notre Dame Press, 1989, 272-302.

[8]Pierre Duhem frequently drew attention to this tension in his works on the history of astronomy, notably in *To Save the Appearances*, transl. E. Doland and C. Maschler, Chicago: University of Chicago Press, 1969. For an extended discussion, see Mc Mullin, "The goals of natural science", sec. 3.

[9]See Mc Mullin, "The explanation of distant action", section 3: "Kepler's dynamics of planetary motion".

[10]See Mc Mullin, The movement of the earth" in *Galileo. Man of Science*, ed. E. Mc Mullin, New York: Basic Books, 1967, 35-42.

[11]*Dialogue on Two Chief World Systems*, transl. Stillman Drake, Berkeley: University of California Press, 19S3, p. 234. See Mc Mullin, "The conception of science in Galileo's work", in *New Perspectives on Galileo*, ed. R. Butts and J. Pitt, Dordrecht: Reidel, 1978, 209-2S7, p. 238.

[12]Mc Mullin, "The movement of the earth", p. 38.

[13]See Mc Mullin, *Newton on Matter and Activity*. Notre Dame: University of Notre Dame Press, 1978, chap. 4.

[14]For a review of some of the main answers, see Mc Mullin, "Conceptions of science in the Scientific Revolution", in *Reappraisals of the Scientific Revolution*, ed. D. Lindberg and R. Westman, Cambridge: Cambridge University Press, 1990, 27-92.

[15]Rene Descartes, *Principles of Philosophy* III, 43; see the translation by V.R. and R.P. Miller, Dordrecht: Reidel, 1983, p. 105.

[16]*Op. cit.*, IV, 205; p. 287. See Mc Mullin, "Conceptions of science", p. 41.

[17]Johannes Kepler, *Apologia Tychonis contra Ursum,* transl. in Nicholas Jardine: *The Birth of History and Philosophy of Science,* Cambridge: Cambridge University Press, 1984, p. 151. See Mc Mullin, "Conceptions of science", p. 58.

[18]Copernicus himself makes this point in the *De Revolutionibus,* I, 10. Kepler amplifies it considerably in the *Mysterium Cosmographicum* of 1596, as well as in the *Apologia.* See Mc Mullin, "Rationality and paradigm-change in science", to appear in a *Festschrift* for Thomas Kuhn, edited by Paul Horwich, M.I.T. Press.

[19]*Mysterium Cosmographicum.* I, 15. See Nicholas Jardine, "The forging of modern realism", *Studies in the History and Philosophy of Science, 10,* 1979, 141-174; p. 157.

[20]Robert Boyle, *New Experiments Physico-Mechanical Touching the Spring of the Air,* 1660. See chap. 1, "Robert Boyle's experiments in pneumatics" (James B. Conant), in *Harvard Case Histories in Experimental Science,* ed. J.B. Conant, Cambridge (Mass.): Harvard University Press, 1957.

[21]Reproduced in R.S. Westfall, "Unpublished Boyle papers relating to scientific method", *Annals of Science, 12,* 1956, 63-73; 103-17.

[22]See "Newton: Deducing from the phenomena" in Mc Mullin, "Conceptions of science", pp. 67-74.

[23]*Opticks,* New York: Dover, 1952, p. 404.

[24]William Whewell, *The Philosophy of the Inductive Sciences Founded upon their History,* London, second edition 1847, New York: Johnson Reprint, 1967, vol. 2, p. 73.

[25]*Op. cit.,* pp. 67-68. See Menachem Fisch, "Whewell's consilience of inductions: An evaluation", *Philosophy of Science, 52,* 1985, 239-255.

[26]See *Scientific Realism,* ed. Jarrett Leplin, Berkeley: University of California Press, 1584, particularly the essays by van Fraassen, Laudan, and Fine.

[27]See, for example, Mc Mullin, "Selective anti-realism", *Philosophical Studies. 58,* 199Q 209-220.

[28]Thomas Kuhn, *The Structure of Scientific Revolutions,* Chicago: University of Chicago Press, 1971, p. 206. Others, like Larry Laudan and Arthur Fine, have pushed the anti-realist implications of Kuhn's work further than Kuhn himself has done.

[29]See, for example, the essays by Richard Boyd and Ernan Mc Mullin in Leplin ed., *Scientific Realism.*

[30]Michael Simon shows how the development of cell biology supports a continuity thesis in *The Matter of Life*, New Haven: Yale University Press, 1971. See also Wesley Salmon, who draws on Jean Perrin's *Les Atomes* (1913) to construct a similar case for atomic theory, in *Scientific Explanation and the Causal Structure of the World*, Princeton: Princeton University Press, 1984, chap. 8.

[31]For case-histories drawn from elementary-particle theory and radio-astronomy respectively, see Andrew Pickering, *Constructing Quarks: A Sociological History of Particle Physics*, Edinburgh University Press, 1984; David Edge and Michael Mulkay, *Astronomy Transformed: The Emergence of Radio Astronomy in Britain*. New York: Wiley, 1976. For particularly strong anti-realist statements, see, for example, Steve Woolgar, *Science: The Very Idea* London: Tavistock, 1988; Harry Collins, *Changing Order: Replication and Induction in Scientific Practice* New Haven: Yale University Press, 1985.

[32]As I have elsewhere suggested. See "Fine-tuning the Universe?", *Proceedings: Conference on Science and Religion* Kentucky State University, to appear.

[33]See Mc Mullin, "How should cosmology relate to theology?", in *The Sciences and Theology in the Twentieth Century*, ed. A.R. Peacocke, Notre Dame: University of Notre Dame Press, 1981, 17-57.

[34]D.W. Sciama, *The Unity of the Universe*, New York: Doubleday, 1961, p. 70.

[35]*Op. cit.* pp. 166-7.

[36]*Op. cit.*, pp. 167, 70.

[37]Another context where they have been notoriously unsuccessful is quantum mechanics. For a discussion, see J.C. Cushing, "Quantum theory and explanatory discourse: An end game for understanding?", *Philosophy of Science. 58*, 1991, 337-358, and the essays in *Philosophical Consequences of Quantum Theory: Reflections on Bell's Theorem*, ed. J. Cushing and E. McMullin, Notre Dame: University of Notre Dame Press, 1989.

[38]See footnote 46 below. It is worth noting that already in 1964 some months before the announcement of the discovery of the cosmic background radiation, Hoyle and Roger Tayler recognized that the "high helium content of cosmic material" could not be explained in terms of ordinary stellar processes but required a "far more dramatic" mode of production in an early "high-temperature high-density" cosmic phase (or else that "massive objects" must have played a larger role in cosmic evolution than had hitherto

been supposed). This was, essentially, to concede that the helium abundance data could not be reconciled with the steady state model though the authors did not say this explicitly. (The mystery of the cosmic helium abundance, *Nature*, 203, 1964, 1108-1110.)

[39]For more detail, see Steven Weinberg, *The First Three Minutes,* New York: Basic Books, 1977, chaps. 3 and 6.

[40]In a recent essay, Ralph Alpher and Robert Herman describe how they were led to this prediction in 1948, and speculate as to why their paper (as well as later papers by the Gamow group developing their conclusion in more detail) did not lead at the time to a search for the cosmic background radiation they had predicted. See "Reflections on early work on 'Big Bang' cosmology", *Physics Today, 41*, Aug. 1988, 24-34.

[41]*Modern Cosmology*. Cambridge: Cambridge University Press, 1971, p. 117.

[42]For a helpful summary, see P.J.E. Peebles et al., "The case for the relativistic hot Big Bang cosmology," *Nature, 352*, 1991, 769-776.

[43]The term 'big bang', though picturesque, was not an especially good choice since it suggests an explosion, in which things are thrown outwards in a pre-existent space. But what the Einstein equation implies is that the space itself suddenly began to expand, carrying all with it, as it were. (Hoyle, who first used the term 'big bang', intended it to be derogatory; see Alpher and Herman, *op. cit.*, p. 24.)

[44]See Alpher and Herman, *op. cit.*, p. 27.

[45]P.J.E.. Peebles, *Physical Cosmology*, Princeton: Princeton University Press, 1971, chap.

[46]Sciama, *Modern Cosmology*, chap. 11.

[47]Deuterium is a fragile isotope of hydrogen; so far as present theory goes, it cannot be produced, nor even survive, within normal stars. All of the deuterium present today (one atom for about every 30,000 atoms of hydrogen) seems likely, therefore, to have come from the original fireball. (One outside possibility has been mentioned, that it might have come from a pregalactic phase of very massive stars.) What makes deuterium a particularly interesting cosmic marker is that, unlike helium, its production in the original fireball would be extremely sensitive to the figure chosen for energy-density. A density high enough to "close" the universe would seem (in the standard model, at least) to yield a zero abundance because all the deuterium produced would be destroyed. The present abundance of deuterium does not, therefore, afford a strong argument for the standard model,

but it may well prove a major clue in determining the initial conditions that would have to be imposed on any big bang type of model. See Joseph Silk, *The Big Bang*, San Francisco: Freeman, 1980, pp. 123-6.

[48]Peebles *et al.*, "Hot big bang cosmology," p. 771.

[49]Peebles et al., give a useful review of the tangled topic of quasar redshifts, in the light of the most recent evidence, *op. cit.*, p. 772.

[50]See the essay by Margaret Geller elsewhere in this volume.

[51]Halton Arp, *Quasars. Redshifts and Controversies*, Berkeley: Interstellar Media, 1989.

[52]Peebles *et al.*, *op. cit.*, pp. 774-5.

[53]Eric Lerner, *The Big Bang Never Happened: A Startling Refutation of the Dominant Theory of the Universe*, New York: Random House, 1991. Lerner's concept of what constitutes "refutation" has come in for a share of (deserved) criticism. The merit, however, even of an overstated and sensationalistic work like this one, is that it leads defenders of the received theory to presuppositions that may have been taken for granted.

[54]This principle is already adumbrated by Descartes in his cosmology. See Mc Mullin, "Fine-tuning the universe?," *Proceedings: Institute for Science and the Liberal Arts*, Kentucky State University, vol. 2, to appear.

[55]C.B. Collins and S.W. Hawking, "Why is the universe isotropic?," *Astrophysical Journal, 180*, 1973, 317-334.

[56]John Leslie in his book *Universes* (London: Routledge; 1989) and in many articles has brought a welcome clarity to this notably tangled topic.

[57]One (the "weak") form of the principle proved to be trivial; the other (the "strong") principle is not really a *principle*, strictly speaking, but is a style of *explanation* of the apparent fine-tuning, involving the presence of humans in the universe as the key factor. See Mc Mullin, "Fine-tuning the universe?."

[58]Alan Guth, "Inflationary universe: A possible solution to the horizon and flatness problems," *Physical Review D, 23*, 1981, 347-356.

[59]See Alan H. Guth and Paul J. Steinhardt, "The inflationary universe", *Scientific American, 215*, 1984 (5), 116-128; Paul Steinhardt, "The current state of the inflationary universe", *Comments Nuclear Particle Physics. 12*, 1984, 273-286; Andrei Linde, "Particle physics and inflationary cosmology," *Physics Today, 4Q* 1987(9), 61-68.

[60]The calculations in regard to primordial nucleosythesis depend initially on the *baryon* density of the universe, omitting other sorts of matter, if such there be. In the standard big bang model the figure for baryon density

that gives the correct abundances for the lightest elements is only about a tenth or less of the critical density marking the boundary between an open and a closed universe. Since the model does not *require* that the density be critical, the most obvious conclusion would be that the density *is* well below the critical level; though the deficit could, of course, be made up by postulating non-baryonic matter, it does not have to be. If the inflationary hypothesis be tacked on, however, the density of the resultant universe *must* be at or very close to the critical value. Hence, on the face of it, this is inconsistent with the figures for the cosmic abundances of the lightest elements, unless one postulates an exotic form of matter making up most of the critical density, leaving the density of baryonic matter still at the smaller value required for nucleosynthesis. What Malaney and Fowler wish to show is that one can manipulate the quark-hadron phase transition (with quark bubbles and other ingenious constructs) in order to get the right elemental abundances, even with the baryonic density as high as the critical value. So there would be no need to postulate exotic matter. (Though a different difficulty still arises: where is the missing *baryonic* matter? Why has it not been observed? The inflationary hypothesis, in this reading, *requires* that the "missing" matter exist; the standard big bang theory, minus the inflationary hypothesis, does not.) Malaney and Fowler's argument is not a positive one for adding on inflation, but a critique of an argument *against* the inflation hypothesis. See Robert A. Malaney and William A. Fowler, "The transformation of matter after the big bang", *American Scientist*, 76, 1988, 472-7.

[61]*Op. cit.*, p. 477.

[62]It is not altogether clear that the inflationary hypothesis *does* eliminate the need for fine-tuning of the initial energy-density. Guth and Steinhardt note that the sort of "slow-rollover" transition required to prevent the inflation from becoming a runaway one once again requires ("unfortunately") a fine-tuning of several of the initial cosmic parameters, notably the energy-density (*op. cit.*, p. 127).

[63]Op. cit., p. 127.

[64]Peebles *et al.*, *op. cit.*, p. 775.

# Questions and Comments

**Fuller:** Again we will start with comments and questions from the panel. Professor Harrison.

**Harrison:** Regrettably scientists have never taken any notice of philosophers. It always seems to scientists that philosophers are struggling to understand what scientists did in the past. Scientists are always sort of struggling with the present and reaching out to ideas of the future. Have philosophers in any way in the 20th century contributed to the advance of science?

**McMullin:** My answer to that is very clear cut. I should hope not. No more than scientists have contributed notably to the furthering of philosophy. In fact, if one had to make comparisons, comparisons are invidious, I suspect that philosophers of quantum mechanics may have contributed a little more to quantum mechanics than any one and quantum mechanics has recently contributed to philosophy. That's my guess. That's invidious so I regard that as not said. But let me make a point that I think is worth underlining here. I think that very often scientists, if you like, will tolerate history of science. After all, they do have a history. Even scientists know that. But, in fact as we heard in many of the lectures here, most scientists, particularly good pedagogs, you heard Margaret earlier today, will know that the presentation of an organic story is part of what it is to make something intelligible and so history of science is clearly important. Philosophy of science is something else again because so frequently in the past it has been arbitrary. In the positivist period in the 1920s with the emphasis on logic and with the kind a priori logical forms, it seemed very often like a rather arid and unconnected construction. I really would urge that that period in the philosophy of science is definitively over. It ended 20 or 30 years ago. And as a matter of fact, what philosophers of science attempt to do today is not to further the work of the scientist except perhaps in some areas like quantum mechanics. Maybe you might argue it there, but basically not. Rather it's to try to understand science itself as a phenomenon. Science is not a body of propositions. It's an activity, the activity of you people. A complicated kind of activity by very complicated people. And so the function of philosophy of science in that sense is a form of history. It's a form of philosophy and incidentally it's also clearly a form of sociology, something that's clearer now than it used to be. And so there is a kind of combined activity. Some people call it science studies where the historical, the philosophical and the sociological come together to understand you very complicated people and how you do what you're doing. Now if one were to be successful in that, the success would be measured not by some sort of furthering of the scientific project. That's unlikely, I think. Scientists themselves are best at that. But rather at a better understanding of the scientific process itself as a process of knowledge, as a process of understanding and there are many intriguing issues.

Let me just mention one. Let me underline one again, Professor Harrison. You, yesterday, expressed a skepticism about what the future may hold in terms of what scientific theories of 50 years hence might be like. One colleague of mine wrote a paper recently called *Extraterrestrial Intelligence* and his argument was that science as it would be constructed by one of Tim Ferris' other civilizations would be of a quite different kind. It wouldn't have, it would be very unlikely to have the atomic table of elements and so forth. If they were bodily different, if they were like lobsters instead of like us, they would have a different etc., etc. You know the stories.

Others have argued in great detail that that's not the case, that they would in fact have a science rather like ours at least in many respects. That debate I think is a very interesting one. It's not going to lead to science being done in a different way, but it obviously has consequences for the kind of topic we're talking about here. None of us, most of us, let's say, will not be around in 50 years. But it's legitimate. But at the same time, I think it's an issue that can be discussed in the light of history. Historians, or philosophers of science, are relying a great deal more on the history of science than they did in the past and I think that's a very healthy development.

**Fowler:** I'd like to quibble with your terminology. To me science activity which you seem, at least in my mind, seem to associate with science, I think of science activity as scientific research. I think of science as the body of knowledge that has been accumulated over the years as a result of scientific research. And I think those two things should be clearly separated. That there is scientific research on the one hand and then there's this great body of knowledge which we call science. I think if you don't do that, a lot of confusion can result.

**McMullin:** That's an interesting issue and let me just make a quick response. You can make a distinction up to a point there and in the past when people talked about science, they talked about it as something in a book as a set of results. But you see, the problem with that, I mean that's all right.

That's what we teach in class, but the problem with that is that there are uneven amounts of credence that you can give to the different claims as you yourself brought out this morning and so the process of assessment is part of this. Let me mention a book that came out just recently written by Peter Galeston at Stanford, now at Harvard, called *How Experiments End*. An elegant book. And what Galeston argues in there is that when you find an experimental result in a book today coming from, say, high energy physics, there's been a lot of arbitration as to whether in fact it's a good result or not. There's been an extraordinarily complex decision process. I would suspect that for example in Dr. Geller's work that the same kind, is it a good result or not, doesn't come easy. And it's that kind of assessment process and the criteria that guided it that philosophers and sociologists have been particularly interested in.

**Morrison:** I was very much impressed by the reflective account which was as much the history of ideas as philosophy going through especially the origin of the cosmological and physical ideas which we were more or less all committed and Craehill was about. But I want to underline the very last of what I took to be the very last topic of Professor McMullin's. It bore very much on what I was trying to say. I didn't have quite enough courage to say it out boldly and now that he's raised the question I think I'm going to do that. It has to do really with my general feeling, you might not agree with me, that the guide of nature intelligibility and of the metaphysical plausibility of pictures is not a very secure guide. The clear example in the past is the work of Newton because the work of Newton who made no hypotheses, who could not describe a mechanical mechanism, a mechanical scheme by which from step by step the gravitating body could pool the thing, was nevertheless I would still say the most seminal of all advances in, arguably in the theory of the cosmos. We couldn't do without it. And the fact that we explain it quite differently now, it really doesn't make much difference. The fact is you use it so much and it's so good. And even the fact that for 100 years it seemed to be sleeping in the sense that we thought it was, Newtonian mechanics, celestial mechanics were kind of little fine details that we would get straight. The perturbation of mercury was the great discovery. And now we see that explosive activity and evolutions of all kind can go on in perfectly gravitational systems. The whole thing is turned upside down, mostly by wider experience and by a lot of computing and we now see we were wrong in considering the boundary condition, as a solar system made it all flat and regular and beautiful and nothing every happens except tiny little things. But if you go to a place that hasn't happened, things are shot out and we don't know what happens. Things collapse and double stars get formed. And you know what it is. It's a paratechnique affair out there.

Well, now something like that I think has happened in a sarabiticious kind of way. It may not have happened, I quite agree. This is not something that is firm. It is something that is definitely on the test. It is on the burner, it's on the rack to be established by the inquisition of the next 15 years, I think. Get out of the world what the true is. And that is the proposal by Guth which carries other things with it of an inflationary explanation for a remarkable quantitative fact. The extraordinary simplicity of the old universe. Not our universe, the old universe.

If what this says is right, I think it throws us necessarily into what I would have to call a brand new state where we will no longer be able to talk about the universe as a whole or the universe as structure. We can only talk about something else. I don't know what to call it yet. It's the local pool or something; it's much bigger than anything we've thought about in the past. But it implies certainly that there's room for theorists—maybe they won't be tested; maybe they won't have good ideas—but there's room for theorists to invent all kinds of things before and around it as Professor Fowler is willing to show what was around it in some vague way. And I think, things about inflation, does think that probably has to be something around it. I'm not sure I'm willing to draw what's around it the way Professor Linde does in an elaborate structure or the way Willie does in a suggestive little one just to show the feeling of the thing. And some are in between. But if this is true, we are now in a position which is a little bit akin to what happened to mechanical knowledge when we began to understand quantum mechanics where there are a lot of questions you are no longer able to answer but it doesn't reduce you to ignorance. It gives you an enormous strength in some of things you can answer. And inflation has a potential for doing that. It may turn out that if inflation turns out very well, you'll have to ask, well is the universe a steady state universe? Maybe, is it a Big Bang universe? Is it both in various places in some other way? Maybe. We have no way of describing, or at least we don't see a way of describing what is inside the local pool which is observable to a degree and which would become more and more observable as time goes on. Which doesn't require any initial conditions except the most minute ones that you can place any kind of space, time and matter and which can probably produce a scalar field and we're left in a very strange quandary which will surely change the science if it is a science forever.

Maybe the only time when it had true high ambitions is the time between 1925 and 1980 when it wasn't so hedged in by wonderful measurements as it is now. So I think we are in a crude state.

**Fuller:** We have a question from the audience. Do you believe personally that there is any reason or benefit for the emergence of a universe with the existence of consciousness?

**McMullin:** A reason for emergence of consciousness? Well, coming at the end of our conference, that reminds me once I was in a conference like this and the chairman said there's only two minutes, but I have a question from the audience and the question was: What is being? I had to find a very short answer to that. There's one sort of banal answer that one can give here. It's obviously not what the questioner has in mind, which is that the theory of evolution, the modified theory of evolution as it has developed today, gives a pretty good account of the appearance of consciousness. It's not perfect any more than the Big Bang theory is. There are a lot of gaps in it. The challenges to it have forced it to in some ways change and revise and so on. But Neo-Darwinian theory has in fact shown its metal, I think, in something of the way the Big Bang model has. And that does give an account of the growth of consciousness over the aeons. There might be another level of question involved in this because those who talked in the 1970s about the anthropic, as I would say, forms of explanation, anthropic principles there are, I think had in mind two things. First of all, that if one has, if one looks at the presence of consciousness in the universe, it seemed to them to demand a long-lived universe in order that evolution could operate. And a long-lived universe in which certain laws of physics notably, certain relationships between the four fundamental forces, would hold. And so the question is how come we have a universe which not only is flat, is fine-tuned as they put it, but also has precisely the laws of physics that would allow for example carbon, would allow planets, would allow galaxies, would allow above all the long life and so forth for evolution. And the answer to that was taken to be the answer or the request for explanation there, was taken to be a request for explanation for the presence of consciousness in the universe. It seems to me that that form of anthropic explanation once again you can go in the many universe direction and say, well of course if many universes exist, then some will in fact almost inevitably have the conditions for consciousness. However, complex they may be.

Another explanation, of course, obviously is the theological explanation. If we see the universe as in some sense part of the unimaginable operation of a transcendent creator, then in fact once again consciousness in the universe will be such as to realize his designs whatever his or her or its designs whatever they may be. But if one pushes, there's an interesting question here that seems to be, that if one takes a third line, which is the line that is implicit I think in the principle, for example, Philip has talked about, which is that the universe had to be of the kind that it is, in a sense if one illuminates contingency in initial conditions, that's the direction you're going in as Dirac did for example, then it sounds as though consciousness is in fact a necessary development of the cosmos and that would be an interesting thought.

**Fowler:** There are perhaps some questions that can never be answered. The question of whether we're the only intelligent species in the universe can never be answered no. It can only be answered positively or just dwindle along in determining. And one of those questions I suspect maybe the issue of whether there is a purpose to nature cast in the following way. The question being, is it more appropriate to regard our curiosity about the universe as ours alone? That is, we're beings who came along perhaps in some accidental way on this planet and we happen to be interested in the wider setting. So it's all our own doing. Or is that interest part of some larger process that we as yet, and perhaps forever, don't know anything about. In other words, if you entertain the possibility that there's some purposeness in nature and I don't know whether there is or not, then there's always the haunting possibility that what one is doing in investigating the universe has more to do with the universe than it has to do with oneself or even one's own species. So, it may be because of questions like that the role of consciousness in the universe may be an unanswerable question, or to put it another way, that it might be one of that list of questions that upon finding oneself in contact with a superior intelligence would be one of the questions they hadn't answered.

**Harrison:** We construct theories of the universe but we find it very difficult to put ourselves and consciousness in those theories but undoubtedly consciousness is a part of reality and we have to remember that we are the universe trying to understand itself. I was very much interested in your views about Newton and things and I know it's not fair because we're not supposed to be talking about it here at this conference, but I'm interested, have you written anything about your views of Bohr's ideas and Schroedinger's and all the recent things that, to me, correspond to Newton?

**McMullin:** I have edited a book recently called philosophical consequences of quantum theory which is an attempt on a part of a number of authors. I called it *Philosophical Consequences of Quantum Theory: Reflections on Bell's Theorem.* And just as we have heard in the course of the last two days, we've caught something of the excitement in cosmology over the last 30 or 40 years. There is a tremendous amount of excitement at the moment over the last, well 10 years, roughly in quantum theory because even though Bell's theorem could in fact have been derived by Heisenberg in the 1930s, in fact, it wasn't. For interesting enough reasons, it wasn't put forward by Bell until the middle '60s and even then its consequences were not fully understood. It had been to some extent foreshadowed by Einstein in the middle '30s but the consequences of that theorem are very far-reaching for all of our principles of natural intelligibility in my view. That is to say, if one considers the kinds of experiments that Bell sketches where you have essentially events occurring at a distance from one another that seem to be correlated, causally correlated, and if you can produce the proof that they can't be causally correlated, at least in any ordinary sense of cause so it has been argued, then you have a real problem. I should think that the next 10 or 20 years is going to be very lively in that area because whatever comes up is quite clear now, will be almost incredible from the point of view of classical physics. I mean quantum theory was already rather incredible but the new forms of action that are called for here are of such a nonintuitive kind. There are several alternatives that are still open but none of them are very attractive ranging from action at a distance which is one, to some very curious form of almost, I hesitate to say the word "ether" but some kind of a very odd medium with very curious properties. There's a great deal of work going on on that and so much work going on and it's quite striking that so far there's simply no agreement at all what inferrences to draw. So that's a very exciting field right now and I should think that at the realm, it's quite striking that just about the time that Guth's work was appearing to make the micro, or the past world, the world of origins, mysterious, so at the level of the very small similar excitement was just beginning to emerge too. I think we have a very exciting period ahead.

# CONVERSATIONS AT NOBEL

## NOBEL DINNER

**Elvee:** So that each of you can hear from our panelists at this dinner, I have proposed to them this proposition. I have asked that they take a little look into the near future—or the far future—and tell us what it is that they're hoping for. So, panelists, what tools are you hoping to get your hands on or to see come into use in the astronomical and related sciences? What theories could be disconfirmed? What projects interest you? Give us a view into your mind as to what you think may be coming along or what you hope may be coming along. I'm going to start with Dr. Morrison.

**Morrison:** I thought it worthwhile to begin with a very proximate, a very soon coming, and a very personal interest of mine just to show what it is that happens to physicists over the years. Many years ago I wrote an extremely optimistic, far too optimistic article on how easy it would be to use gamma rays as a window on the universe—how they would bring us direct information of nuclear reactions and relativistic particles and all sorts of wonders which you could not find any other way. There was some truth in that, but, as usual, the theorists circled, enormously cavalier about how easy it is to do things. Because for them it's just a matter of pushing the pencil and the paper a few times, but for somebody else, it's a long, difficult process.

And so it turned out to be a long, difficult process. While gamma ray astronomy has had some wonderful successes, largely due to a satellite that was a German-Netherlands cooperation about 8 or 10 years ago, a really substantial look into the gamma ray world is being taken for the first time by a satellite that was launched in May by NASA called GRO for Gamma Ray Observatory. It has the wonderful attribute that its antenna was stuck and sticking antennas are a terrible bane of people who fly vehicles, but the brave crew of the shuttle got into their extravehicular activity suits and went out and released this antenna, saving us $300 million and an awful lot of hopes that had gone for 8 or 10 years in that satellite.

So, it works like a charm. It went through six weeks of testing and now is going through a schedule. It's looking in the sky, spot after spot, constellation after constellation, wonder after wonder. It's already found 117 gamma ray bursts and nobody quite understands where they come from. And at the end of next week, it is going to look at the most wonderful object which specially delights me. I spent a year trying to figure out something about the gamma rays that came out of it and about ten years thinking about what to do with that. At the end of next week I think the GRO will point toward 3CC273 and I'm hoping to have something come in.

Now, this is very nice. It took us 35 years to get to this state so it's a kind of realization which is very pleasing. And, of course, I'm quite sure what the result will be—something ambiguous. It will be interesting, but ambiguous, that will turn us back again to rethink and recalculate and increase the sensitivity and do the rest of the things that people have to do in order to puzzle out the world which is always one step, fortunately, out of our grasp.

It's not ten steps out, because then we'd give up. It's not within our grasp, because then we'd be finished. But it's just out there at the edge and that's very desirable and I think that's the way gamma rays are going to be.

I want to talk about a quite different point which I feel is important to say here. This campus has been so hustling and so friendly and so interesting. Science and astronomy, in particular, has its fruit. The wonderful things we've heard the last few days and the arguments and the intellectual level of our inquiry and the sense of the giant tradition, ancient tradition, that leads us on, are all very important. But they represent the fruit of the study.

But we all know that fruits need seeds. In fact, they are nothing but the vehicle of seeds. The whole thing is a cycle. The genetics of the trade is what concerns me very much at the present time.

Who are going to be the astronomers of the 1990s and the 2000s? Not us, except maybe for Willie (Fowler). A few will continue, but we need a full parade of new people, always new people, always new ideas, always new proposals, always new audacity, always new soberness.

I think that one of the most important tasks the astronomers face and scientists as a whole is to recruit more widely in the structure of our educated people than we ever have before in science. The most exemplary point that I could make, and the most obvious one, is that half the population, somewhat the smarter half of the population, is rather poorly represented in astronomy. Imagine doubling the force. What progress we'll make toward solving these problems if we learn how to induce young women students to become astronomers.

And the smaller groups in our population who have been underrepresented in these learned professions, we need them in astronomy. I hope they'll not all be attracted by worldly successes and Porsches. If they want to do something which I think has a more permanent value internally and what will matter to the future of our species, then this is the thing to do—get into astronomy.

Of course, more broadly speaking, any one of the intellectual activities in science is open to everyone but we need to make the sciences still more welcoming and still more available. And I hope that will happen. It's beginning to happen. In another ten or twenty years we'll see it fully at high surge.

**McMullin:** We've been asked a difficult question this evening--look ahead ten years and give us your wish list. I suppose in line with the themes of the conference, one-half of my wish list would dwell on the themes of mysterious action in the universe at the realm of the very small and the very distant--the point I made earlier today.

We're just at the threshold, it seems to me, both in quantum theory and cosmology, of new and very revolutionary forms of understanding that are so far from the intuitive, so far from the ordinary everyday world that they turn that world upside down. Older forms of thought were based on the assumption that the rest was much like the familiar. Science, and not just science, extended other forms of thought, extended outwards by analogy from the familiar. That was part of what it was all about.

That doesn't work now. The quantum world is very unfamiliar. The world of the inflation, the inflation and the first fraction of a second of the universe is very unfamiliar, very unintuitive and the forms of action that occurred then are still far beyond our understanding. They lead us, I hope, to deep humility in the face of that.

That leads me to the other half of my wish list. I want to raise an ending, a theme that we have not, in fact, touched on at the conference here but I think it's appropriate.

I speak here in a Swedish Lutheran college. I come from a Catholic university and I hope you'll forgive me for a moment touching on a theological issue. If one goes back in the most ancient western religious traditions and, in particular, the Hebrew tradition, one finds one or other of the great literary works of the ancient world. The Book of Isaiah, one of the greatest of all pieces of literature, or the Book of Psalms. One finds throughout those wonderful works of literature a cosmological imagination. If you haven't read Isaiah or Job or Psalms lately, go back and see how much the cosmological infuse their imagination. They speak of the stars, they speak of the mountains, they speak of the firmament. They touch constantly on cosmological themes.

The Hebrews were excited by that. Not in the way the Greeks were and for different reasons but they were excited by it. And for them the theological was very much associated with the cosmological.

For reasons we're familiar with and which would take a long time to go into, a great separation has occurred. The theological imagination today which is still very active is very removed from the cosmological. There has been a divide there. And part of the reason for that was evident at this conference.

The cosmological is very difficult. It is very complex and above all it's very dismaying. The universe that we heard of so effectively in the last couple of days is a vast universe, a universe of many regions, of regions far beyond our reach and far beyond our imagining. If the inflationary model is correct, it's even so incredibly beyond imagining that we quail in the face of it.

But if that's the case, and if the theologian is serious about his claim, that God is the creator of that universe, then that being, whoever, whatever it is, is so far beyond our imagining that theologians have to start afresh.

I think the challenge to the theologian today is even more exciting than it is for the cosmologist or the quantum theorist. Because whichever mode of understanding he or she follows, whether he thinks of God as a transcendent being, responsible for the fact that there is a universe rather than no universe, or whether he finds a universe, God has to be in some sense part of that universe, no matter which mode of understanding he follows. The inflation that has happened in cosmology over the last thirty years, may lead within the lifetime of some people here to an equivalent inflation in theology.

**Ferris:** The wish I have for the future is not involved with equipment or tools, but with the state of our education and with the conveyance of our culture to the future generations. Science is at the center of our culture. It's what this society does best, as far as I can see. It's worth doing. It's important, it's vital and it's not a set of beliefs or dogma, but a way of looking at the world that is liberating to the person who has it and that's what's important to me.

I don't care very much whether the United States happens to be the country that builds the next fastest airplane or the next bullet train, or something of that sort. And I suppose I could live with it if our grand tradition of 20th century physics and astronomy were passed on to some other culture, as long as it was being done somewhere.

But, it's the absence of freedom that I'm concerned about. Because people that don't read and don't learn how to think are not free people. I'm finding increasingly that among our students that there are fewer and fewer of them who are free in this sense.

So, my hope for the future is that we can recover from this recent period in particular. Phil (Morrison) made a mention of material goods and Porsches and so forth. I've gone through a string of Porsches so I can't very well decry driving a Porsche as evidence of a decadent lifestyle or, for that matter, as evidence of an intellectual property.

But it does seem to be a fact that we have just gone through a lamentable epoch, which I regard as one of the most disgraceful in our national history. We have, during this period, spent the inheritance of our descendants as if there was no tomorrow and during this same period our intellectual life has suffered its most precipitous decline. Historically, as a nation, we have been the best off intellectually during periods of economic hardship—the 1930s, for instance, when some of our finest intellectuals went through their formative years.

I fear such a period of economic difficulty is ahead, but the one bright spot I see is that perhaps many of our younger people will recover a sense of values and priorities. I might say I've seen exampled at this institution such a sense and that is the reason that I am so happy to have been allowed to be here and so grateful to our hosts for their kind hospitality during the two days I've been privileged to visit the college.

**Harrison:** We've been asked to saying something meaningful. While visiting with Timothy Ferris about what we were going to say, I took out this slip of paper (which he and I had received), this kind of fortune cookie which asked us about the tools and the projects of the next decade. I couldn't help thinking about that story of the guy who was walking down a New York street talking to a woman about his fantasy. In his fantasy he is walking with Leonardo da Vinci, showing him all the helicopters, planes and buildings, telling him what modern society is like and then Leonardo da Vinci says, "How wonderful. You must explain to me how it all works." And the guy says, "That's where my fantasy comes to an end."

You see, we live in a society in which increasingly we are understanding less and less of the technical and scientific underpinnings of this great society. More and more, this is becoming a scientifically illiterate society in defiance of what made it so great.

Young people are turning away from science into the legal studies, into business school and, therefore, the problem is who is going to replace the Willie Fowlers, Philip Morrisons, Margaret Gellers, Ernan McMullins, and Timothy Ferrises. Where are the replacements coming from?

In our graduate school in the physical sciences 80% of our students are Asians. Most of them will return to the Pacific rim. Who is going to take over in this country? Where will the best graduate schools in the world in the physical sciences be? Who is going to be teaching in those schools in twenty years time?

Somehow, the whole business of this society has become geared to realizing immediate profit and that philosophy is now applying in the selection of careers. What is going to make me a millionaire the quickest so I can retire--sick of what I've chosen--at the age of forty-five? The choice to follow some career that is intellectual and will be spiritually or intellectually satisfying, independent of how little it might pay you, is not being made.

So, when I'm asked what are the tools and projects I'd like to see in the next decade, I answer that I'd like to see human resources. I'd like to see more and more people, our young people, turning to science.

**Geller:** I'd like to do two things. First, I'd like to make a selfish wish and then I'd like to tell you a brief anecdote.

To make the wish, I'd like to quote one of the famous philosophers of our age--RobotMan. In a cartoon not long ago, RobotMan comes along and says to his friend, "You know, I have a complete map of the universe." And his friend says, "A complete map?" And RobotMan says, "Yeh, I'd show it to you, but it's really hard trying to fold it back up." So, I wish I could be that person with that complete map of the universe.

Then I'd like to share with you a short story. A week ago a father and his high school-aged son came to my office ostensibly for professional advice. The son came in and said to me, "Professor Geller, you know, I've always wanted to be an astronomer but my father doesn't like this idea because he thinks it's impractical. Is it impractical?"

I looked at him and said, "Well, yes! He's right. It's impractical but it doesn't matter." The father looked at me and asked various questions about salary and where people worked and so on. It went on and on. And finally I said to the son, "You know, the thing to do when you want something like this is follow your dreams."

And I felt like saying to the father, "You're really lucky to have a son who has these dreams."

**Fowler:** I have just celebrated my 80th birthday and if you think I'm going to talk about the future! You're crazy. Don't make me talk about the past either. No, I decided that I would hear enough serious things from the rest of the members of this distinguished group and so the only laughs I got at my talk today was my story, so I'm going to tell a story. The first thing I'm going to do is quote Mark Twain. As I think it says in my biography, I'm a great admirer of Mark Twain. I think I've read all of his books. But the quotation from Mark Twain that I like best is—"There's something wonderful about science. One gets such wholesale returns of conjecture from such a small investment in fact." Boy, he didn't know how right he was. And then I would like to talk about, or just say one last thing, a story about a Nobel Prize winner who I met in England many years ago, Sir John Cockcroft. I'd always been a great admirer of Cockcroft because it was Cockcroft and Walton using a little accelerator in the cavindish lab who showed that you could disintegrate nuclei which has been my business all my life, by accelerating charged particles and letting them slam into stationary targets and studying the reaction products where produced. So I've spent many years in England. My wife and I go every year either to England or Scotland and I got to know Cockcroft well enough that at some time or other in my acquaintance with him in Cambridge I was displeased with something that was going on in the university. I forget what it was. So I went to Cockcroft to see if I could get him to do something about it. And I explained in a sentence something about what I wanted and he said, "yes." And I went on and with another sentence and he said, "yes." And I went on and on and on and every time I made a statement about what I was interested in, he said, "yes." And two months went by and nothing happened. So I went to a friend of mine who also knew Cockcroft and I said, "something's wrong. When I was in to see Sir John about this matter in which I had great interest, every time I put a proposition to him, he said yes and it's been two months now and nothing has happened." "Oh," he said, "when Cockcroft says yes, he means, yes, I've heard you!"

Well, I've had a wonderful time here. I want to thank all of you people who have been so nice to us. I'm going to take off tomorrow for California and get started doing some work again.